Electrical Characteristics of Transmission Lines

Wolfgang Hilberg

Charakteristische Größen
elektrischer Leitungen

Eine Einführung in die Berechnung von
Wellenwiderständen, Kapazitäts- und Induktivitätsbelägen
homogener elektrischer Leitungen vom Zylinder- und Kegeltyp

VERLAG BERLINER UNION GMBH. STUTTGART
VERLAG W. KOHLHAMMER GMBH.
STUTTGART BERLIN KÖLN MAINZ

Wolfgang Hilberg

Electrical Characteristics of Transmission Lines

An Introduction to the calculation of characteristic impedances
and specific capacity and inductance of homogeneous
cylindrical and conical electrical transmission lines

ARTECH HOUSE BOOKS

Translated from the German by:

Chester E. Claff, Jr., Ph.D.
Linguistic Systems, Inc.
Cambridge, Ma. 02139

Copyright © 1979
ARTECH HOUSE, INC.
610 Washington Street
Dedham, Massachusetts 02026
Printed and bound in the United States of America
Library of Congress Card Catalog Number: 79-23940
Standard Book Number: 0-89006-081-9

Table of Contents

Preface

Foreword

Introduction

Part A

THE GENERAL CALCULATION METHOD AND THE
DETERMINATION OF THE CHARACTERISTICS OF
A SERIES OF TRANSMISSION LINES

Chapter I

Chapter II

Chapter III

Chapter IV

Chapter V

Chapter VI

Chapter VII

Chapter VIII

Chapter IX

Part B

Part C

Part D

Preface

The properties of transmission-line conductors is a subject which is interesting and informative from various viewpoints. On one hand, there are a large variety of cross-sectional shapes which have practical utility. On the other hand, every shape exemplifies a mathematical function suited for its exact or approximate evaluation. Both of these objectives are served by the fascinating collection of examples presented herein.

The usual simplification is achieved by the assumption of quasi-static fields, which means that all or most of the energy is within a cross section so small that the interaction of electric and magnetic fields can be ignored. The further assumption of perfect conductors and free-space dielectric enable the same formula to give not only wave resistance but also the inductance and capacitance per unit length. Therefore all three scales appear in each set of graphs.

The pair of cylindrical conductors, which is typical of a long line, is evaluated for 60 shapes. In addition to the usual round wires and planar strips, there are arc-strips of mainly theoretical interest.

Most of the formulas are stated for analysis (rather than synthesis). The magnetic-field loss by the skin effect can therefore be evaluated by analytical or numerical differentiation, except for the strips, which have no thickness.

The corresponding pair of biconical conductors is described for each cylindrical pair. Some of the biconical pairs are interesting for launching a spherical wave, though the radiation is beyond the scope of this presentation.

One feature is the evaluation of every shape in terms of simple functions. Some shapes are formulated exactly. The others are approximated, usually by two overlapping formulas for lower and higher wave resistance.

Another feature is the collection of graphs for all shapes, with the three scales of wave resistance, inductance and capacitance.

The author received his degree of Dr.-Ing. in the Technical University of Darmstadt in 1963. After some years of experience with

Telefunken, he became Professor and Dean at Darmstadt. Most recently, he is Head of the Institute of "Datentecknik" (Digital Techniques). His many technical articles cover a variety of specialties, of which the subject of this book is just one. This presentation is a testimonial to his curiosity and ingenuity in developing a set of interesting concepts and related formulas.

This excellent translation from the original German becomes a valuable addition to the English-language library.

Harold A. Wheeler
Greenlawn, N.Y.
September, 1979

Foreword

The main features of the following observations were reported in two internal reports of AEG-Telefunken in early 1967. It was soon recommended to the author that he make the thoughts contained in them available to a still larger circle of readers, since the vivid and largely unknown calculation methods and the great number of computational results would certainly be very useful to many engineers. However, it was clear to the author that the reports would have to be considerably improved and broadened for this purpose. Unfortunately it has taken a rather long time to bring the manuscript to its present state. This was primarily the result of the demands of many other projects on the author, but probably also because the methods had already in fact been worked out and only the laborious work of completion and verification with the use of numerous examples remained. However, since the author set out again and again to complete the work, and the favorable reception of the circulating internal reports seemed to confirm that this was no useless matter, a definite conclusion was eventually reached. (It is planned to continue the work somewhat further in the future, wherein the problem of lines subject to losses and inhomogeneous lines will be considered in a similar comprehensive manner as is the case here for homogeneous, loss-free lines.) For all that, the long completion time had something in its favor. To be specific, it turned out clearly that the interest in the characteristics of high-frequency transmission lines and particularly of striplines has not diminished (one need only observe the increasing number of publications over the years in the literature references). Also, from another viewpoint, the requirement for more comprehensive working literature has not been satisfied in the meantime. Besides the well-known article by Ch. A. Hachemeister in 1958, which reviews a certain class of cylindrical transmission lines by conventional methods, nothing similar has been published up to the present time. Other authors, such as Grivet (1970), limit themselves generally to an introductory discussion of the general relationships. Thus, the present paper might be able to contribute a little to filling the existing gaps. The goal is as much to give the engineer suitable working documents, as to facilitate entry for students into the problem area of calculations for transmission lines.

Ulm, summer, 1971   W. Hilberg

Introduction

The characteristic impedance of transmission lines has always been of great interest to both practitioners and theoreticians. The first needs the knowledge of its numerical value in order to choose suitable currents and voltages and to be able to match lines, and the other as a rule, finds the relatively simply formulated problem to be difficult enough for him to look for generally usable solutions for special transmission line cross sections not yet studied. As a result of this, over a period of time, starting with completely simple line cross sections such as that of the coaxial cable, characteristic impedance formulas have already been derived for the widest variety of line cross sections. As a rule, one proceeds to idealize the curves of the line cross section geometrically, and also assumes loss-free conductor materials, and then calculates an analytical expression for the characteristic impedance with these assumptions. The derivation can take a very wide variety of forms. Furthermore, one can concentrate on the calculation of the so-called "exact" formula or can be satisfied with approximations. Good approximations are completely adequate in general for the practitioner since "exact" solutions with few exceptions can be illustrated only by non-elementary functions, wherein it also must be considered that the geometrical and electrical idealizations in actuality also often lead to errors, which can no longer be completely ignored. On the other hand, from the theoretical viewpoint, strictly derived complete solutions are naturally worth striving for. Both viewpoints, therefore, have their justification, and tables of characteristic impedances of common transmission lines therefore frequently contain both "exact" formulas and approximation formulas.

The way in which formulas have been collected into tables in the past is now a clear sign of the fact that the calculation processes in general are difficult and that no uniform process for the calculation of the formulas is really available. This can be seen purely formally in the fact that the existing tables often contain formulas with various constants or with constants not labelled precisely, and that the precision obtainable is almost never indicated for approximation formulas.

It will now be shown below that the situation can be fundamentally changed. By a method based primarily on stereographic projection and combined with the known methods of the corresponding diagrams, the reflections, and the utilization of properties of complementary lines and most of the characteristic impedance formulas contained in the known tables can be systematically obtained simply and intuitively from the formula of a single line. It is essential here that each transformation undertaken changes nothing of the precision of a solution, i.e., an "exact" formula remains "exact", and a solution with a specific error deviation retains this deviation precisely. Now, in order not to be committed from the beginning to exact formulas or approximations (in the deviations), so-called characteristics are determined in Part A, the use of which makes possible excellent approximations from non-elementary functions and also the exact formula. What characteristic impedance formulas are thereby arrived at is shown in Part B, where approximations from elementary functions are determined with precisely defined maximum deviations for the previously treated transmission lines, and are presented in tables. At the same time, this constitutes a substantially improved survey of characteristic impedances of cylindrical and conical transmission lines in comparison with the existing compilations (cylinder and cone are to be understood here with their general meanings, not only as circular cylinders and circular cones!). With respect to a convenient practical application, where initially only the order of magnitude is of pressing interest, the tabulated formulas of Part B are also illustrated graphically in Part C. A detailed list of references in Part D, where somewhat obscure transmission lines can be found treated in some papers, or where it can be determined roughly with what intensity many problems of transmission line calculation have been treated, completes this work.

The purpose of the first section A is to show how the tabulated formulas in Part B were obtained and, more importantly, to illustrate in full detail how the formulas for other cases not treated here can be arrived at. The stereographic projection, of course, is a very far-reaching aid for the determination of characteristic impedances, which has not previously been recognized in the literature. In fact, all parallel lines considered here can basically be calculated even without the assistance of the stereographic projection, but the use of the conical lines with the use of the stereographic projection greatly facilitates the treatment and thereby presents itself directly for systematic developments. That is to say, it will be shown that a single conical line is always associated with a large number of parallel lines; i.e., all parallel lines resulting from one another by linear transformations have a single equivalent conical line. It therefore turns out that all types of transmission lines within a linear relationship can be taken in at a glance, as it were. On the other hand, the scope of the linear relationship can very easily be exceeded by such simple operations as reflections, formation of complementary lines, changes of scale with subsequent projection, and naturally also by introduction of ordinary non-linear transformation. By a constant interchange between conical lines and cylindrical lines, the range of known lines can be constantly broadened. Considerable advantages are produced in comparison with other known calculation procedures. As is well known, the use of a non-analytical-numerical process even limits one completely to the calculation of the specific system, whose field in addition should be limited in space. Related lines here are often recognizable only with difficulty. Difficulties arise also with certain idealized lines, since, for example, with infinitely thin strips or points, the field strength exceeds all limits at some positions, etc. In an analytical calculation, such difficulties are circumvented as a rule. The transmission lines thus chosen for reasons of clarity, and the lines with unrestricted field space should, therefore, always be calculated analytically, or at least transformed by interposed transformations into forms most suitable for a non-analytical-numerical calculation.

In comparison with the many known works for the calculation of transmission lines in which use has also been made of the conformal mapping, a fundamental difference also exists in the following paper inasmuch as the attempt has probably never been made before to make use of transformations as systematically as is done here. The important point in the following discussions in Part A will be found in the treatment of idealized striplines, the dimensioning of which is of great current importance. However, the derivations might first be preceded by some more general considerations, in which the problem will be described in somewhat more detail. The reader acquainted with these problems can skip over them.

Chapter I

THE GENERAL CALCULATION METHOD
AND THE DETERMINATION OF THE
CHARACTERISTICS OF A SERIES OF
TRANSMISSION LINES

Statement of the Problem and Preliminary Considerations

Elementary transmission line theory [5] shows that the ratio of voltage to current produces the so-called characteristic impedance Z, which is calculated for loss-free transmission lines with the use of the inductance per length L' and the capacity per length C', by:

$$Z = \sqrt{\frac{L'}{C'}} \tag{1}$$

With the use of the speed of propagation of waves along the line,

$$v = 1/\sqrt{L'C'} = 1/\sqrt{\mu\epsilon}$$

and with the charge per unit length $Q' = C' \cdot U$ (1) can also be written:

$$Z = \frac{\sqrt{\mu\epsilon}}{C'} = \frac{L'}{\sqrt{\mu\epsilon}} = \frac{\sqrt{\mu\epsilon}}{Q'} \cdot U \tag{2}$$

With this, the calculation of the characteristic value Z for the wave propagation along loss-free transmission lines is reduced to the calculation of the capacity per unit length (or of the inductance per unit length), i.e., an electrostatic problem.

It is also well known that conformal mapping can be used to advantage in the calculation of characteristic impedances. For this purpose, the cross-sectional plane is defined as the complex z plane and a transformation into a u-v plane is undertaken with an analytical function f(z), i.e.,

$$w = u + iv = f(z) = f(x + iy).\tag{3}$$

With such mapping, curves in the x-y system are transformed into the u-v system isogonally in detail, but with a certain distortion of scale.

Therefore, if two conductors of a cylindrical transmission line (parallel line) are given the contours K_1 and K_2 in cross section, for example, see Figure 1, and if the calculation of the capacity in the x-y coordinate system is too difficult, then it is attempted to find a coordinate transformation or a conformal representation in which the contours K_1 and K_2 coincide with the constant coordinates in the u-v plane or until known and easily calculable cross sections are produced. It is not difficult to show [7] that in every conformal mapping, the capacity and therefore also the characteristic impedance, remains unchanged. Therefore, if only these values are of interest, as in the following, then as many transformations as desired can be carried out successively without worry, and there is no need to be concerned about the varying pattern of field lines between the conductors in each case.

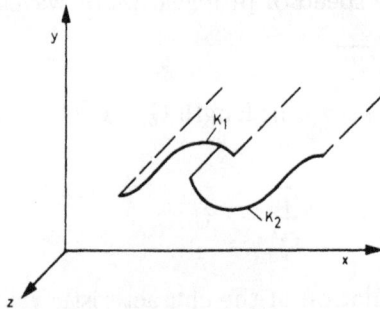

Figure 1

Finally, it should again be pointed out as a reminder, that all circles again are changed over to circles only in a linear transformation.

A few further comments concerning the relationships between planar and spherical wave propagation are probably necessary for

the following. Let there be a conical transmission line consisting
of two conductors of arbitrarily good conductivity, whose
surfaces are formed by the rays of an arbitrary angular shape
projecting outward from the origin (feed point). Figure 2 shows
some examples. If a spherical wave now propagates outward
from the feed point between these conical conductors, then this
can take place in the form of a TEM wave, as is well known.
Surfaces of identical phase are spherical surfaces in which lie
the transverse field parameters E_t and H_t.

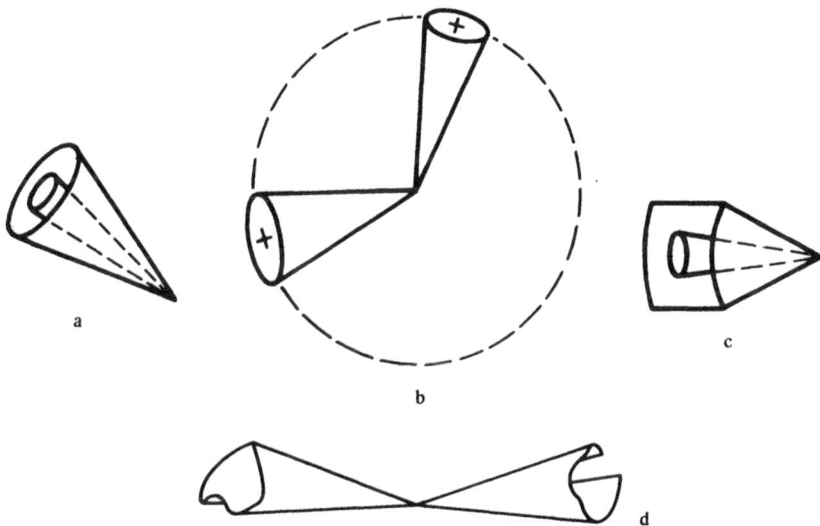

Figure 2

The wave equation

$$\Delta\Phi + k^2 \Phi = 0, \quad k^2 = \omega^2 \mu\epsilon \tag{4}$$

is valid for it, wherein Φ is a symbol for E_t or H_t.

The variation of these values with the radius can now be expressed
by the equation:

$$\Phi = V(\vartheta, \varphi)\frac{e^{-ikr}}{r}. \tag{5}$$

The operator $\Delta\Phi$ in spherical coordinates, r, Θ, Φ, now reads [8]

$$\Delta\Phi = \frac{\partial^2 \Phi}{\partial r^2} + \frac{2\partial\Phi}{r\partial r} + \frac{1}{r^2 \sin^2 \vartheta} \frac{\partial^2 \Phi}{\partial\varphi^2} + \frac{1}{r^2} \frac{\partial^2 \Phi}{\partial\vartheta^2} \tag{6}$$

$$+ \frac{1}{r^2} \cot \vartheta \frac{\partial\Phi}{\partial\vartheta}$$

If equation (5) is now introduced into this equation, after insertion into the wave equation (4), the following equation results:

$$\sin \vartheta \frac{\partial}{\partial\vartheta} \left(\sin \vartheta \frac{\partial V}{\partial\vartheta} \right) + \frac{\partial^2 V}{\partial\varphi^2} = 0. \tag{7}$$

The variation with r has dropped out. Therefore, only the solution of the (now static) potential problem on a spherical surface remains. This problem in turn can easily be converted into a planar problem with the use of a specific transformation. Let:

$$\rho = \tan \vartheta/2. \tag{8}$$

Then:

$$\frac{\partial\rho}{\partial\vartheta} = \frac{1+\rho^2}{2}, \tag{9a}$$

and

$$\frac{\partial V}{\partial\vartheta} = \frac{\partial V}{\partial\rho} \frac{\partial\rho}{\partial\vartheta} = \frac{\partial V}{\partial\rho} \frac{1+\rho^2}{2}. \tag{9b}$$

Together with the known theorem

$$\sin \vartheta = \frac{2 \tan \vartheta/2}{1 + \tan^2 \vartheta/2} = \frac{2\rho}{1+\rho^2} \tag{10}$$

this can now be inserted into (7) whereby the following is obtained:

$$\frac{2\rho}{1+\rho^2} \frac{\partial}{\partial\vartheta} \left[\frac{2\rho}{1+\rho^2} \frac{\partial V}{\partial\rho} \frac{1+\rho^2}{2} \right] + \frac{\partial^2 V}{\partial\varphi^2} = 0. \tag{11}$$

By using (9a) and simplifying the expression in the brackets, the following is finally obtained:

$$\rho \frac{\partial}{\partial\rho} \left(\rho \frac{\partial V}{\partial\rho} \right) + \frac{\partial^2 V}{\partial\varphi^2} = 0. \tag{12}$$

However, this is exactly the form of the Laplace Equation $\Delta V = 0$ in planar polar coordinates [8]. Furthermore, since the propagation of TEM waves along cylindrical transmission lines amounts directly to the solution of the same equation, there are obviously very close relationships between cylindrical and conical transmission lines. This is the fundamental reason why alternate use can advantageously be made below of spherical and planar wave propagations.

Chapter II

The Stereographic Projection

If it is desired in geography to depict the surface of the earth (globe) on a plane (flat map), it can be done with the use of the so-called stereographic projection. Its invention is attributed to Hipparch, in approximately 140 BC (sometimes also attributed to Ptolemy, who lived at the same time). The projection is accomplished in the manner sketched in Figure 3. A flat surface F is tangent to the sphere, e.g., at the position of the North Pole. Rays are now assumed to project out from the South Pole, projecting the point P or curves on the surface of the sphere onto the plane. Such a projection is isogonal and in particular, depicts circles on the surface of the sphere as circles or straight lines on the plane. In Figure 3, it can be shown with elementary geometrical considerations through isosceles triangles, that the peripheral angle is equal to half of the central angle ϑ. If the diameter of the sphere is now assumed to be 1, for simplicity, the result is the projection equation

$$\rho = \tan \vartheta/2. \tag{13}$$

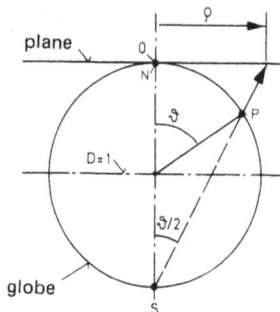

Figure 3

The stereographic projection, therefore, as is established by comparison with (8) is precisely the transformation needed mathematically to bring the spatial problem back to a planar problem [3], [4]. As is well known, the stereographic projection has also in fact been intensively investigated in the theory of functions. In this case, one speaks of Riemann mapping or the mapping of the complex number sphere (actually the surface of the sphere) on the complex plane [20]. It corresponds to the conformal and specifically to a linear representation [29], it is also somewhat more precisely designated as a circular relationship of the second type [31] and has a great deal of similarity to the transformation by reciprocal radii in the plane. For this reason, all equations derived for conformal mapping in the plane are also valid for the stereographic projection. In particular, it can be concluded from this, that the electrical characteristics of conical transmission lines such as characteristic impedances, capacities, and inductances, are identical with those of the cylindrical transmission lines which are obtained by projection.

For the sake of completeness, it should also be noted in conclusion that except for the preparation of maps, to a great extent stereographic projection has previously been used practically exclusively in crystallography, where a flat representation is prepared from the space structure; for example, refer to [27].

Chapter III

Complementary and Reflected Transmission Lines

In addition, we shall need one more important relationship between the characteristic impedances of complementary transmission lines. A transmission line is complementary if electrical and magnetic walls of the system are interchanged exactly with one another, as shown in Figure 4a, b for a coplanar cylindrical line and in Figure 4c, d, e for a coplanar conical line, by way of examples.

Such complementary systems have characteristic impedances Z_i and Z_{ii} which obey the relationship:

$$Z_I \cdot Z_{II} = \frac{\eta^2}{4} .$$ (14a)

For brevity, we set:

$$\eta = \sqrt{\frac{\mu}{\epsilon}} = \sqrt{\frac{\mu_r \mu_o}{\epsilon_r \epsilon_o}} \text{ with } \sqrt{\frac{\mu_o}{\epsilon_o}} = 120\pi \text{ Ohm} \approx 377 \text{ Ohm.}$$ (15)

For the calculation of areas in which the H-lines do not close, as in Figure 4f, g or in the upper hemiplane of Figure 4a, b, for example, instead of (14a) the following relationship is applicable for the corresponding Z_{TI} and Z_{TII}

$$Z_{TI} \cdot Z_{TII} = \eta^2 = \frac{\mu}{\epsilon} .$$ (14b)

(A detailed derivation of these equations is provided in Section VIII 5). The reflection of transmission lines will be used very frequently in the following discussion. New transmission line cross sections are formed by this well-known elementary operation, wherein the characteristic impedance changes merely by the factors of 2 or 1/2.

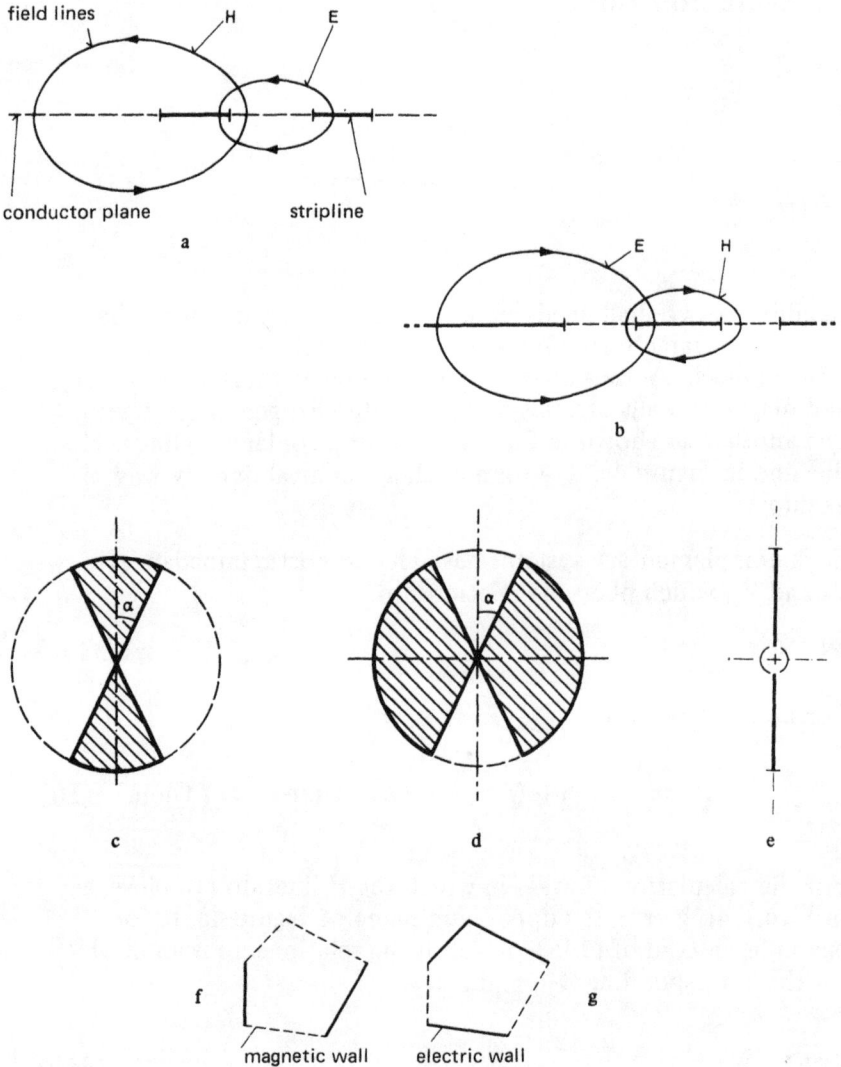

Figure 4

Chapter IV

Auxiliary Equation

An equation very useful later can be obtained as follows: a circular cone with the aperture angle β as in Figure 5 is to be represented stereographically on a plane. Let the cone axis have the angle α towards the north-south axis and on the plane, let ρ be the smallest distance and ρ' the greatest distance of the projected circle from the origin. According to (13), the following mapping equations then apply:

$$\rho = \tan \frac{\alpha - \beta}{2} \text{ and } \rho' = \tan \frac{\alpha + \beta}{2}, \tag{16}$$

and therefore, converted

$$2 \text{ arc tan } \rho = \alpha - \beta$$
$$2 \text{ arc tan } \rho' = \alpha + \beta. \tag{17}$$

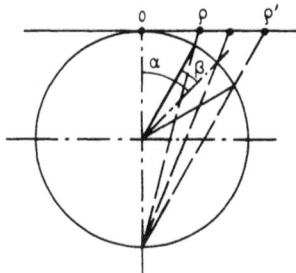

Figure 5

By addition of the two equations, β is eliminated, and with the use of the addition theorem:

$$\text{arc tan } u \pm \text{arc tan } v = \text{arc tan } \frac{u \pm v}{1 \mp uv} \tag{18}$$

one obtains:

$$\frac{\rho + \rho'}{1 - \rho\rho'} = \tan \alpha. \tag{19}$$

This equation will be used frequently in the following for $\alpha = 90°$. For this, it then follows:

$$\rho\rho' = 1 \tag{20}$$

and, since ρ, ρ' are standardized values ($\rho = r/a$; $\rho' = r'/a$):

$$r \cdot r' = a^2. \tag{21}$$

Chapter V

Characteristic Impedances Expressed by Elementary Functions (Solid Conductors)

1. Coaxial cable with elliptical cross section

For most of the following discussions, we need only a single direct calculation of the characteristic impedance of one transmission line. This will now be carried out. Consider the transformation:

$$w = \text{ar cosh } \frac{z}{a}, \qquad (22)$$

wherein a is a constant factor.

This can be transformed:

$$\frac{z}{a} = \cosh w = \cosh (u + iv) = \cosh u \cos v + i \sinh u \sin v \qquad (23)$$

$$= \frac{x}{a} + i \frac{y}{a}.$$

Therefore:

$$x = a \cosh u \cos v, \quad y = a \sinh u \sin v. \qquad (24)$$

If v is eliminated from these two equations, there results:

$$\frac{x^2}{a^2 \cosh^2 u} + \frac{y^2}{a^2 \sinh^2 u} = 1,$$

Form: $\dfrac{x^2}{A^2} + \dfrac{y^2}{B^2} = 1,$ \qquad (25) wherein A is the large semiaxis and B is the small semiaxis. If u is eliminated on the other hand, then:

$$\frac{x^2}{a^2 \cos^2 v} - \frac{y^2}{a^2 \sin^2 v} = 1, \text{ Form: } \frac{x^2}{A^2} - \frac{y^2}{B^2} = 1. \tag{26}$$

For constant values of u, (25) describes confocal ellipses, since the eccentricity e here is:

$$e^2 = A^2 - B^2 = a^2 (\cosh^2 u - \sinh^2 u) = a^2. \tag{27}$$

And for constant values of v, (26) describes confocal hyperbolas, since here:

$$e^2 = A^2 + B^2 = a^2 (\cos^2 v + \sin^2 v) = a^2. \tag{28a}$$

Figure 6b shows the curves in the x-y plane. Each such line of constant value of u or v can now be interpreted as the tracing of a conductor surface, e.g., the lines u_1 and u_2. The potential difference is now equal to the applied voltage, $u_1 - u_2 = U$, the line charge Q' results from $Q' = \int \vartheta ds = \epsilon \int E \, ds$, which can also be written in the form $E = \frac{\partial u}{\partial n} = \frac{\partial v}{\partial s}$ as a consequence of the Cauchy-Riemann equations applied to $Q' = \epsilon \oint dv = \epsilon \, 2\pi$ and which by insertion in (2) leads to the equation

$$Z = \sqrt{\frac{\mu}{\epsilon}} \, \frac{u_1 - u_2}{2\pi}. \tag{28b}$$

Therewith, the characteristic impedance between the two elliptical conductors, obtained from (22) and with (15), is immediately:

$$Z = \frac{\eta}{2\pi} \left(\text{ar cosh } \frac{A_1}{a} - \text{ar cosh } \frac{A_2}{a} \right), \tag{29}$$

(in Table 1, No. 2)

wherein A_1, A_2 signify the two large semiaxes. For small eccentricities a, the arguments become very large, and with the help of the transformation

$$\text{ar cosh } x = \ln (x + \sqrt{x^2 - 1}), \tag{30}$$

it can be recognized that (29) then furnishes the characteristic impedance for the coaxial cable with circular cross section; see Figure 6a:

$$Z = \frac{\eta}{2\pi} \ln \frac{r'}{r} \; .$$

(31)

(in Table I, No. 1)

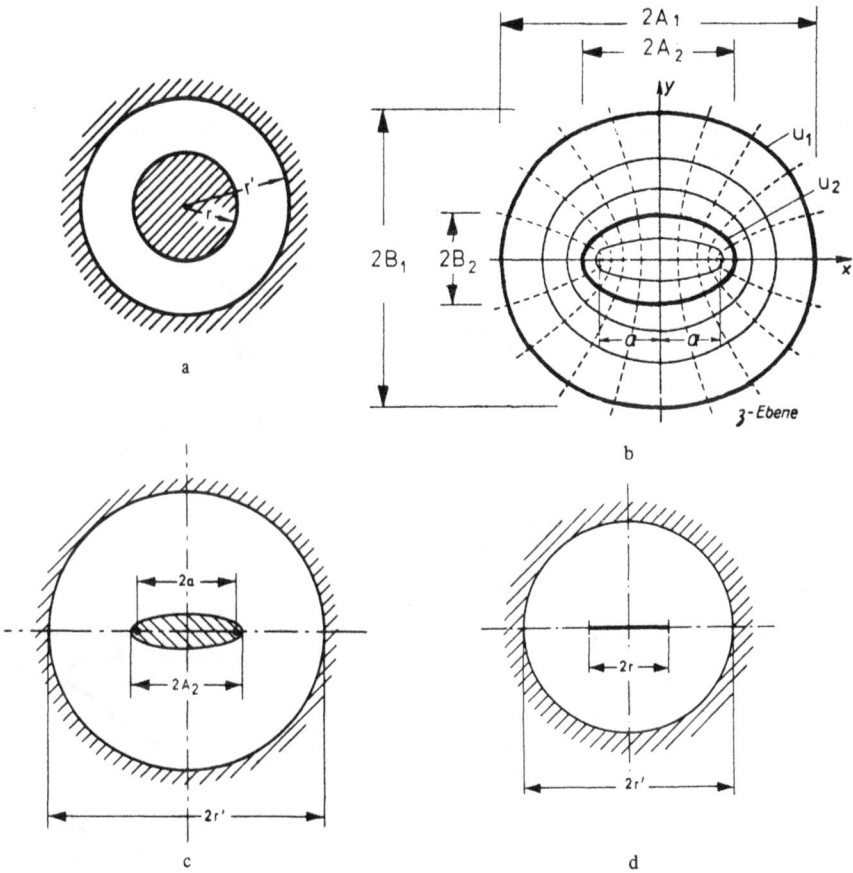

Figure 6

With increasing diameter of the outer conductor, the values of the two semiaxes A_1 and B_1 in Figure 6b approach one another because of (27), whereupon the outer conductor becomes approximately circular. For a small ellipse pursuant to Figure 6c concentric in a circle of radius r', the characteristic impedance obtained to a good approximation is therefore:

$$Z \approx \frac{\eta}{2\pi} \left(\ln \frac{2r'}{a} - \text{ar cosh} \frac{A_2}{a} \right) , \; A_1 \approx r'.$$

(32)

Finally, the small semiaxis B_2 can then be allowed to approach 0, from which because of (27), A_2 becomes equal to a. For a narrow thin strip thus formed in a circle pursuant to Figure 6b, the characteristic impedance then becomes:

$$Z \approx \frac{\eta}{2\pi} \ln \frac{2r'}{r} .\tag{33}$$

The narrower the strip, the better the approximation, and it becomes the exact value for $r \to 0$.

2. *The coaxial circular cone line*

The calculation of the characteristic impedance of a coaxial circular cone transmission line from a coaxial circular cylindrical line, is considered first as the simplest example of the application of stereographic projection. The necessary mapping is shown in cross section in Figure 7a, where for better clarity here and in the following, the tangential plane is always folded along its trace and one half is folded upward into the plane of the drawing.

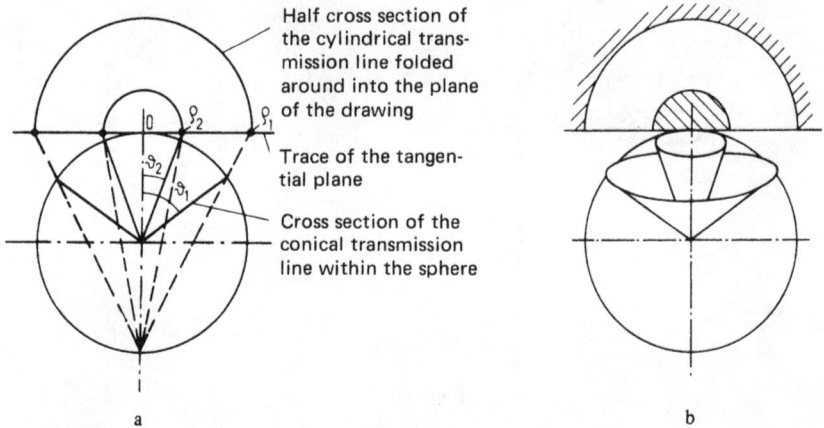

Half cross section of the cylindrical transmission line folded around into the plane of the drawing

Trace of the tangential plane

Cross section of the conical transmission line within the sphere

Figure 7

If the four points $\rho_1, \rho_2, -\rho_1, -\rho_2$ are transferred to the sphere, because of the circular faithfulness of the projection, the circular cone can immediately be drawn in. Figure 7b should make this somewhat clearer in perspective. Later on, however, this will not generally be done. From the characteristic impedance for the coaxial circular cylindrical cable (31) written as:

$$Z = \frac{\eta}{2\pi} \ln \frac{\rho_1}{\rho_2} \qquad (34)$$

where ρ is to be interpreted as the standardized parameter $\rho = r/a$, with $\rho = \tan /2$ and with the designations of Figure 7, by insertion into (34), one immediately obtains the characteristic impedance:

$$Z = \frac{\eta}{2\pi} \ln \frac{\tan \vartheta_1/2}{\tan \vartheta_2/2} . \qquad (35a)$$

(in Table II, No. 1)

This equation is apparently strictly valid for all values of the angles ϑ_1 and ϑ_2. For the three cases in which the angle ϑ_1 is a right angle, an obtuse angle (with the complementary angle $\Theta = \pi - \vartheta_1$) and $\vartheta_2 = \Theta$, we can write:

$$Z = \frac{\eta}{2\pi} \ln \cot \vartheta_2/2, \quad \vartheta_1 = \pi/2, \qquad (35b)$$

(in Table II, No. 2)

$$Z = \frac{\eta}{2\pi} \ln \frac{\cot \Theta/2}{\tan \vartheta_2/2}, \quad \vartheta_1 = \pi - \Theta \geqslant \pi/2, \qquad (35c)$$

$$Z = \frac{\eta}{2\pi} \ln \cot^2 \Theta/2, \quad \vartheta_2 = \Theta. \qquad (35d)$$

If the common conical axis in Figure 7 is now turned by an arbitrary angle, the conical transmission lines and their characteristics naturally do not change. However, various new conductor cross sections are now obtained in the projection onto the plane. Thus, for example, as shown in Figures 8, 9, and 10, the series progresses to an eccentric cable, two round conductors opposite one another, and one round conductor above a conductive plane. Finally at the conclusion of this Section V, the circular conical transmission line with different axial directions, of interest in antenna engineering, as shown in Figure 11, will be studied. The few examples whose characteristic impedances can be completely described by elementary functions are probably thereby completely covered. Their calculation is not difficult, but strangely somewhat more involved than that for the stripline in Section VI below.

3. The Eccentric Cable

In Figure 8a:

$$\rho_1 = \tan\frac{1}{2}\left(\frac{\pi}{2}-\vartheta_1\right) \text{ and } \rho_2 = \tan\frac{1}{2}\left(\frac{\pi}{2}-\vartheta_2\right). \tag{36}$$

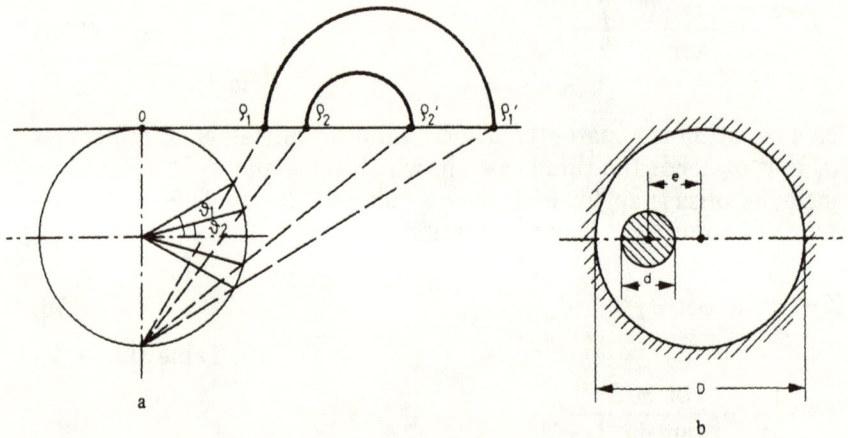

Figure 8

If this is solved for ϑ_1 and ϑ_2, and the equations are inserted into (35a) we obtain:

$$Z = \frac{\eta}{2\pi}\ln\frac{\tan\left(\frac{\pi}{4}-\text{arc tan }\rho_1\right)}{\tan\left(\frac{\pi}{4}-\text{arc tan }\rho_2\right)} \tag{37}$$

With the addition theorem:

$$\tan(\alpha\pm\beta) = \frac{\tan\alpha\pm\tan\beta}{1\mp\tan\alpha\tan\beta} \tag{38}$$

this can be transformed into:

$$Z = \frac{\eta}{2\pi}\ln\frac{(1-\rho_1)(1+\rho_2)}{(1+\rho_1)(1-\rho_2)} \tag{39}$$

With the use of (21), it follows that:

$$Z = \frac{\eta}{2\pi}\ln\frac{(a-r_1)(a+r_2)}{(a+r_1)(a-r_2)}. \tag{40}$$

In analogy with the following equations:

$$y = \ln x , \quad -y = -\ln x \quad \frac{1}{2}(e^y + e^{-y}) = \tag{41}$$

$$\cosh y = \frac{1}{2}(x + \frac{1}{x}), \text{ or } y = \ln x = \text{ar } \cosh \frac{1}{2}(x + \frac{1}{x})$$

it is now desirable to undertake another small transformation, from which it follows that:

$$Z = \frac{\eta}{2\pi} \text{ ar } \cosh \frac{1}{2} \left[\frac{(a-r_1)(a+r_2)}{(a+r_1)(a-r_2)} + \frac{(a+r_1)(a-r_2)}{(a-r_1)(a+r_2)} \right]. \tag{42}$$

The terms in the brackets can be simplified by multiplying out and simplifying, observing that $r_1 r_1' = r_2 r_2' = a^2$, pursuant to (21), from which we obtain:

$$Z = \frac{\eta}{2\pi} \text{ ar } \cosh \frac{(r_1'+r_1)(r_2'+r_2) - 4r_1 r_1'}{(r_1'-r_1)(r_2'-r_2)}. \tag{43}$$

If it is desired to express the characteristic impedance using the usual parameters D, d, e, in accordance with Figure 8b, the following equations apply:

$$D = r_1' - r_1$$

$$d = r_2' - r_2$$

$$e = \frac{r_1'+r_1}{2} - \frac{r_2'+r_2}{2} \tag{44}$$

$$r_1 r_1' = r_2 r_2'.$$

The parameters r_1, r_1', r_2, r_2' here can easily be replaced by D, d, e; of these, let us indicate only the following:

$$r_1' = \frac{1}{2e}\left(e + \frac{D}{2} - \frac{d}{2}\right)\left(e + \frac{D}{2} + \frac{d}{2}\right) \tag{45}$$

$$r_1 = r_1' - D.$$

Some computation work is saved if the parameters r_1, r_1' are written as follows:

$$2r_1' = (r_1' + r_1) + (r_1' - r_1) \tag{46}$$
$$2r_1 = (r_1' + r_1) - (r_1' - r_1),$$

from which the product $4r_1 r_1'$ which appears in (43) turns out to be:

$$4r_1 r_1' = (r_1' + r_1)^2 - (r_1' - r_1)^2. \tag{47}$$

If the terms with the factor $(r_1' + r_1)$ in the numerator of (43), which are known from (45), are combined, then only members remain which can be expressed directly by (44), and it finally follows that:

$$Z = \frac{\eta}{2\pi} \text{ ar cosh } \frac{D^2 + d^2 - 4e^2}{2Dd} \tag{48}$$
$$\text{(in Table I, No. 3)}$$

4. Two Round Wires

In Figure 9a:

$$\rho_2 = \tan \frac{1}{2}\left(\frac{\pi}{2} - \vartheta_2\right) \text{ and } \rho_1 = \tan \frac{1}{2}\left(\vartheta_1 - \frac{\pi}{2}\right). \tag{49}$$

Solving for ϑ_1 and ϑ_2 and insertion in (35a) gives:

$$Z = \frac{\eta}{2\pi} \ln \frac{\tan\left(\frac{\pi}{4} + \text{arc tan } \rho_1\right)}{\tan\left(\frac{\pi}{4} - \text{arc tan } \rho_2\right)}. \tag{50}$$

With the addition theorem in (38), this can be converted to:

$$Z = \frac{\eta}{2\pi} \ln \frac{(\rho_1 + 1)(1 + \rho_2)}{(1 - \rho_1)(1 - \rho_2)}. \tag{51}$$

By resolving the scale and conversion as in (41), it follows that:

$$Z = \frac{\eta}{2\pi} \text{ ar cosh } \frac{1}{2}\left[\frac{(a + r_1)(a + r_2)}{(a - r_1)(a - r_2)} + \frac{(a - r_1)(a - r_2)}{(a + r_1)(a + r_2)}\right]. \tag{52}$$

The expression in the brackets can be combined and simplified by the use of

$$Z = \frac{\eta}{2\pi} \text{ ar cosh } \frac{(r_1' + r_1)(r_2' + r_2) + 4r_1 i_1'}{(r_1' - r_1)(r_2' - r_2)}. \tag{53}$$

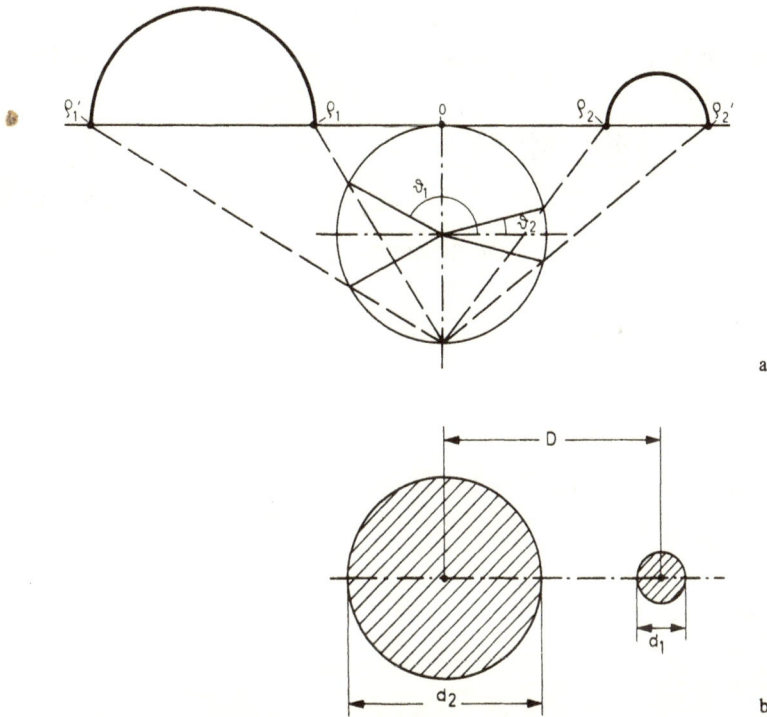

Figure 9

The product $4r_1 r_1'$ can again be replaced by (47) and a conversion can be carried out into the usual parameters D, d_1, d_2, in accordance with Figure 9b. The following equations apply:

$$D = \frac{r_1 + r_1'}{2} + \frac{r_2 + r_2'}{2}$$

$$d_1 = r_1' - r_1 \qquad\qquad (54)$$

$$d_2 = r_2' - r_2$$

$$r_1 r_1' = r_2 r_2',$$

from which is obtained, for example:

$$r_1 = \frac{1}{2D}\left(D - \frac{d_1}{2} - \frac{d_2}{2}\right)\left(D - \frac{d_1}{2} + \frac{d_2}{2}\right) \qquad (55)$$

$$r_1' = r_1 + d.$$

With $(r_1 + r_1')$ from this equation and the equations in (54), all expressions in (53) can now be expressed by D, d_1, d_2, and finally it follows that:

$$Z = \frac{\eta}{2\pi} \text{ ar cosh } \frac{4D^2 - d_1^2 - d_2^2}{2d_1 d_2} \qquad (56)$$
(in Table I, No. 4)

For circular wires of the same diameter: $d_1 = d_2 = d$, from which it follows that:

$$Z = \frac{\eta}{2\pi} \text{ ar cosh } \frac{2D^2 - d^2}{d^2} . \qquad (57)$$

With the well-known equation:

$$2 \text{ ar cosh } u = \text{ar cosh } (2u^2 - 1) \qquad (58)$$

this becomes:

$$Z = \frac{\eta}{\pi} \text{ ar cosh } \frac{D}{d} . \qquad (59)$$
(in Table I, No. 5)

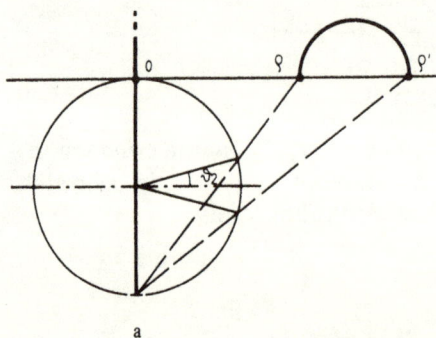

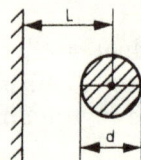

Figure 10

The case of the circular conductor above a conductive plane could be treated by the transformation in Figure 10a. However, it is simpler to obtain the value of interest from the result for two identical circular conductors, since the plane of symmetry in that case constitutes a plane of constant potential, and can therefore be replaced by a conductive plane. The characteristic impedance of a circular conductor opposite this plane then amounts to one-half of the characteristic impedance between the two circular conductors, from which in Figure 10b with L = D/2, it follows that:

$$Z = \frac{\eta}{2\pi} \text{ ar cosh } \frac{2L}{d} \tag{60}$$

(in Table I, No. 6)

5. Circular cone line with different axial directions

By mapping on the sphere the cross section of two circular wires of different diameters, whose characteristic impedance is now known, the characteristic impedance of two cones as in Figure 11, whose aperture angles are ϑ_1 and ϑ_2 and whose cone axes describe an angle α, can now also be calculated easily.

From (53) the characteristic impedance of the circular wire transmission line, if the standardized parameter ρ is used instead of r, amounts to:

$$Z = \frac{\eta}{2\pi} \text{ ar cosh } \frac{(\rho_1'+\rho_1)(\rho_2'+\rho_2) + 4\rho_1\rho_1'}{(\rho_1'-\rho_1)(\rho_2'-\rho_2)} . \tag{61}$$

From Figure 11a, it follows that

$$\rho_1 = \tan\frac{\alpha_1-\vartheta_1}{2}, \qquad \rho_1' = \tan\frac{\alpha_1+\vartheta_1}{2},$$

$$\rho_2 = \tan\frac{\alpha_2-\vartheta_2}{2}, \qquad \rho_2' = \tan\frac{\alpha_2+\vartheta_2}{2}. \tag{62}$$

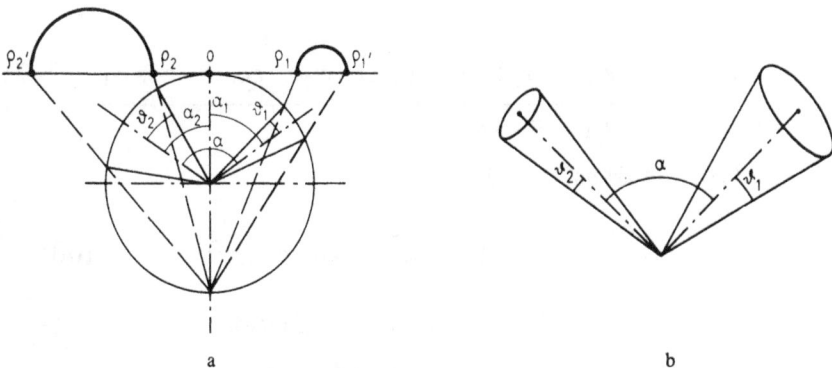

Figure 11

Also, by suitable choice of the zero point or by turning the conical transmission line, the following equation is produced:

$$\rho_1 \cdot \rho_1' = \rho_2 \cdot \rho_2' . \tag{63}$$

With the addition theorem:

$$\tan \alpha \pm \tan \beta = \frac{\sin(\alpha \pm \beta)}{\cos \alpha \, \cos \beta} \tag{64}$$

from (61) by insertion of (62), now becomes:

$$Z = \frac{\eta}{2\pi} \text{ ar cosh} \tag{65}$$

$$\sin \alpha_1 \, \sin \alpha_2 + 4 \sin ((\alpha_1 - \vartheta_1)/2) \sin ((\alpha_1 + \vartheta_1)/2) \times$$

$$\cos ((\alpha_2 + \vartheta_2)/2) \cos ((\alpha_2 - \vartheta_2)/2) \times \frac{1}{\sin \vartheta_1 \, \sin \vartheta_2}$$

With the further addition theorems:

$$\sin (\alpha \pm \beta) = \sin \alpha \cos \beta \pm \cos \alpha \sin \beta \tag{66}$$

$$\cos (\alpha \pm \beta) = \cos \alpha \cos \beta \mp \sin \alpha \sin \beta$$

the numerator of the argument is further transformed, so that we have:

$$Z = \frac{\eta}{2\pi} \text{ ar cosh} \tag{67}$$

$$\frac{\cos \vartheta_1 \cos \vartheta_2 - \cos (\alpha_1 + \alpha_2) + \cos \vartheta_1 \cos \alpha_2 - \cos \vartheta_2 \cos \alpha_1}{\sin \vartheta_1 \, \sin \vartheta_2} .$$

From (63) or the transformed form:

$$\tan \frac{\alpha_1 + \vartheta_1}{2} \tan \frac{\alpha_1 - \vartheta_1}{2} = \tan \frac{\alpha_2 + \vartheta_2}{2} \tan \frac{\alpha_2 - \vartheta_2}{2} \tag{68}$$

by the use of the following transformation equation:

$$\tan (\alpha + \beta) \cdot \tan (\alpha \quad \beta) = \frac{\sin (\alpha + \beta) \sin (\alpha - \beta)}{\cos (\alpha + \beta) \cos (\alpha \quad \beta)} \tag{69}$$

$$= \frac{\cos^2 \beta - \cos^2 \alpha}{\cos^2 \beta - \sin^2 \alpha}$$

it now also follows that:

$$\frac{\cos \vartheta_1}{\cos \vartheta_2} = \frac{\cos \alpha_1}{\cos \alpha_2}, \tag{70}$$

so that the two last terms in the numerator of (67) drop out. With this, we have the result:

$$Z = \frac{\eta}{2\pi} \text{ ar cosh } \frac{\cos \vartheta_1 \cos \vartheta_2 - \cos \alpha}{\sin \vartheta_1 \sin \vartheta_2}. \tag{71}$$
(in Table II, No. 3)

Chapter VI

Characteristic Inpedances Expressed by
Elementary, Arbitrarily Precise, Selectable
Approximation Equations for Non-
Elementary Functions (Striplines)

1. Double cone strip reference system

In (33) the characteristic impedance of a narrow thin strip opposed
to an enveloping elliptical, nearly circular cylinder, was calculated.
This system can now be depicted stereographically on the sphere,
in order to obtain the characteristic values of the conical system of
Figure 12. For this figure, it follows that:

$$\rho = \frac{r}{a} = \tan \frac{\alpha_2}{2}, \quad \rho' = \frac{r'}{a} = \tan \frac{\alpha_1}{2}. \tag{72}$$

Insertion in (33) then provides the characteristic impedance of the
narrow conical strip (e.g., of a thin metallic strip) opposed to the
coaxial, nearly circular cone:

$$Z \approx \frac{\eta}{2\pi} \ln \left(2 \frac{\tan \alpha_1 /2}{\tan \alpha_2 /2} \right), \alpha_2 \ll \alpha_1 . \tag{73}$$
(in Table II, No. 4)

If the angle α_1 is allowed to become $\pi/2$, then to a good approximation there exists the case of the conical strip perpendicular to
a conductive plane, for which the characteristic impedance therefore reads:

$$Z \approx \frac{\eta}{2\pi} \ln \ (2 \cot \alpha_2 /2), \alpha_2 \ll \pi/2. \tag{74}$$
(in Table II, No. 5)

For reasons of symmetry, the characteristic impedance Z_h of the
narrow double conical strip of Figure 4c assumes twice the value:

$$Z_h = \frac{\eta}{\pi} \ln \left(2 \cot \alpha/2\right), \, 0 \leqslant \alpha \leqslant \frac{\pi}{4}. \tag{75}$$
(in Table II, No. 6)

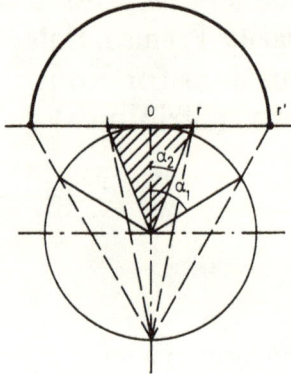

(The index h is intended to indicate a high resistance and the index 1 a low one). The complementary systems, however, are very wide double conical strips almost in contact with one another, as in Figure 4d. As a result of (14) the characteristic impedance Z_1 of the complementary transmission line can be obtained from (75) with the same precision as that of the narrow conical strip:

$$Z_1 = \frac{\eta^2}{4 Z_h} = \frac{\pi \eta}{4} \bigg/ \ln \left(2 \cot \left(\frac{\pi}{4} - \frac{\alpha}{2}\right)\right), \, \frac{\pi}{4} \leqslant \alpha \leqslant \frac{\pi}{2}. \tag{76}$$
(in Table II, No. 6)

Finally, the maximum divergences for the boundaries between the areas of validity occur at $\alpha = \pi/4$. They can be immediately stated [1] if it is considered that because of (14), the exact value designated as Z_m, turns out to be

$$Z_m = \eta/2, \quad \alpha = \pi/4. \tag{77}$$

The maximum divergences, therefore, are approximately $+2.37 \cdot 10^{-3}$ for Z_h and approximately $-2.36 \cdot 10^{-3}$ for Z_1.

For most applications, such precision is completely adequate. However, in [1, 10, 28], algorithms were also derived with which

such approximations can be improved to any desired degree. In
combination with these algorithms, the approximations (75) and
(76) are accordingly completely equivalent to the so-called "exact"
representation by elliptical integrals.

For better comprehension, let us compare the different repre-
sentations of the characteristic impedance of double conical strips.
It can be shown that the characteristic impedance can be described by elliptical integrals in the form [1]:

$$Z = \frac{\eta}{2} \frac{K(k)}{K(k')}, \quad K(k) = \int_0^{\pi/2} \frac{d\varphi}{\sqrt{1-k^2 \sin^2 \varphi}},$$

$$K(k') = \int_0^{\pi/2} \frac{d\varphi}{\sqrt{1-k'^2 \sin^2 \varphi}}, \tag{78}$$

$$k = \cos \alpha, \quad k' = \sin \alpha.$$

The two-component expression (75, 76), with an error smaller
than $2.4 \cdot 10^{-3}$, after slight transformation, is:

$$Z_h = \frac{\eta}{\pi} \ln\ 2\left(\frac{1 + \cos \alpha}{\sin \alpha}\right), \quad \text{for } \frac{\eta}{4} \leqslant Z \leqslant \infty \text{ and } 0 \leqslant \alpha \leqslant \frac{\pi}{4}$$

$$Z_1 = \frac{\pi\eta}{4} /\ln\left(2\ \frac{1 + \sin \alpha}{\cos \alpha}\right), \quad \text{for } 0 \leqslant Z \leqslant \frac{\eta}{4} \text{ and } \frac{\pi}{4} \leqslant \alpha \leqslant \frac{\pi}{2}. \tag{79}$$

(in Table II, No. 6)

By use of the algorithm [1, 19, 28] for the same area of validity,
a two-component expression is obtained with an error smaller than
$3 \cdot 10^{-6}$

$$Z_h = \frac{\eta}{4\pi} \ln\left(2\ \frac{1 + \sqrt{\cos \alpha}}{1 - \sqrt{\cos \alpha}}\right)$$

$$Z_1 = \eta \frac{\pi}{4} /\ln\left(2\ \frac{1 + \sqrt{\sin \alpha}}{1 - \sqrt{\sin \alpha}}\right). \tag{80}$$

Another two-component expression with a maximum error now lowered to less than $4 \cdot 10^{-12}$ is:

$$Z_h = \frac{\eta}{8\pi} \ln \left(2 \frac{\sqrt{1 + \cos \alpha} + \sqrt[4]{4 \cos \alpha}}{\sqrt{1 + \cos \alpha} - \sqrt[4]{4 \cos \alpha}} \right)$$

$$Z_1 = \eta \frac{\pi}{2} / \ln \left(2 \frac{\sqrt{1 + \sin \alpha} + \sqrt[4]{4 \sin \alpha}}{\sqrt{1 + \sin \alpha} - \sqrt[4]{4 \sin \alpha}} \right) . \tag{81}$$

This series of improvements can be continued as far as desired. These examples are probably sufficient to illustrate the assertion that such approximate expressions can meet all requirements both practically and theoretically.

If the characteristic impedances of other transmission lines are obtained by transformations from the impedance of symmetrical double conical strips, then it suffices to follow the change of the parameter α and to take into consideration also a numerical factor λ appearing in addition to the parameter η, and resulting from possible reflections. As an example, we should like to proceed from the approximation (79) in the following, and to refer to it in the form:

$$Z_h = \lambda \frac{\eta}{\pi} \ln [2k_A]$$

$$Z_1 = \lambda \frac{\pi \eta}{4} / \ln [2k'_A], \tag{82}$$

wherein, because of the easily understood relationship [1]:

$$k'_A = \frac{k_A + 1}{k_A - 1} \tag{83}$$

only k_A is to be considered as an independent parameter appearing instead of the parameter α, which can be expressed somewhat less favorably. However, the formulas with elliptical integrals (78) can also be obtained at any time from the following results (or the reverse), since the modulus k follows from k_A in the following manner [1]:

$$k = \frac{k_A{}^2 - 1}{k_A{}^2 + 1} .$$

(84)

2. *Conical strip and circular cone, coaxial*

The characteristic impedance formulas of the conical strip perpendicular to a conductive plane are obtained by halving the values from (75, 76), i.e.,:

$$k_A = \cot \alpha/2, \quad \lambda = \frac{1}{2} .$$

(85)
(in Table II, No. 5)

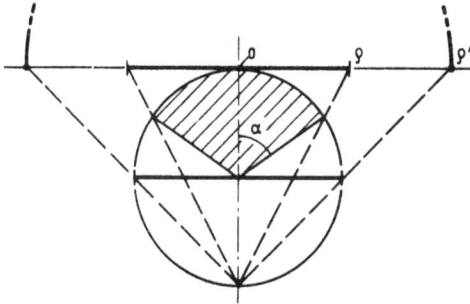

Figure 13

By projection of this system onto the plane as in Figure 13, the strip turns into an enveloping cylinder. It follows from this, that:

$$\rho' = \frac{r'}{a} = \tan \pi/4 = 1, \quad \rho = \frac{r}{a} =$$

(86)

$$\tan \alpha/2, \text{ bzw. } \frac{a}{r} = \cot \frac{\alpha}{2} = \frac{r'}{r} .$$

By insertion into (85), the new parameters are obtained:

$$k_A = \frac{r'}{r}, \quad \lambda = \frac{1}{2} .$$

(87)
(in Table I, No. 8)

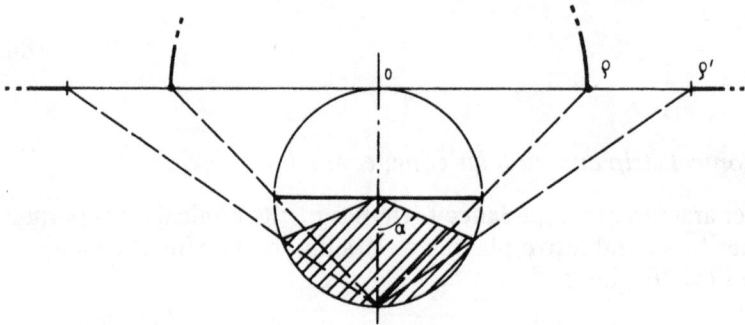

Figure 14

The conical strip standing perpendicular to a plane can naturally also be projected into other positions on the plane. If this is done for expediency as in Figure 14, a circular conductor is obtained in the opening in a conductive plane. From this, it follows that:

$$\rho = \frac{r}{a} = \tan \pi/4 = 1, \quad \rho' = \frac{r'}{a} = \frac{r'}{r} = \tan \left(\frac{\pi}{2} - \frac{\alpha}{2} \right) . \tag{88}$$

Solved for $\alpha/2$ and inserted into (85) with the use of the addition theorem, we have:

$$\cot (\alpha \pm \beta) = \frac{\cot \alpha \cot \beta \mp 1}{\cot \beta \pm \cot \alpha} \tag{89}$$

or simply with the use of $\tan \left(\frac{\pi}{2} - \frac{\alpha}{2} \right) = \cot \frac{\alpha}{2}$ it follows that

$$k_A = \frac{r'}{r}, \quad \lambda = \frac{1}{2}. \tag{90}$$
$$\text{(in Table I, No. 45)}$$

If the conical system of Figure 14 is rotated around the axis perpendicular to the plane of the drawing by the angle $\pi/2$, then a projection onto the plane as in Figure 15a produces a strip perpendicular to a plane, Figure 15b. It follows that:

$$\rho = \frac{r}{a} = \sqrt{\frac{r}{r'}} = \tan \frac{1}{2} \left(\frac{\pi}{2} - \alpha \right) . \tag{91}$$

Solved for $\alpha/2$ and inserted into (85) with the use of the addition theorem (89), it follows that:

$$k_A = \frac{\sqrt{r'/r + 1}}{\sqrt{r'/r - 1}}, \quad \lambda = \frac{1}{2}. \tag{92}$$

(in Table I, No. 11)

It turns out from this, for reasons of symmetry, that a stripline of equal width lying in a plane as in Figure 15c with $D' - 2r'$, $D = 2r$, has twice the value of this characteristic impedance, or:

$$k_A = \frac{\sqrt{D'/D + 1}}{\sqrt{D'/D - 1}}, \quad \lambda = 1. \tag{93}$$

(in Table I, No. 9)

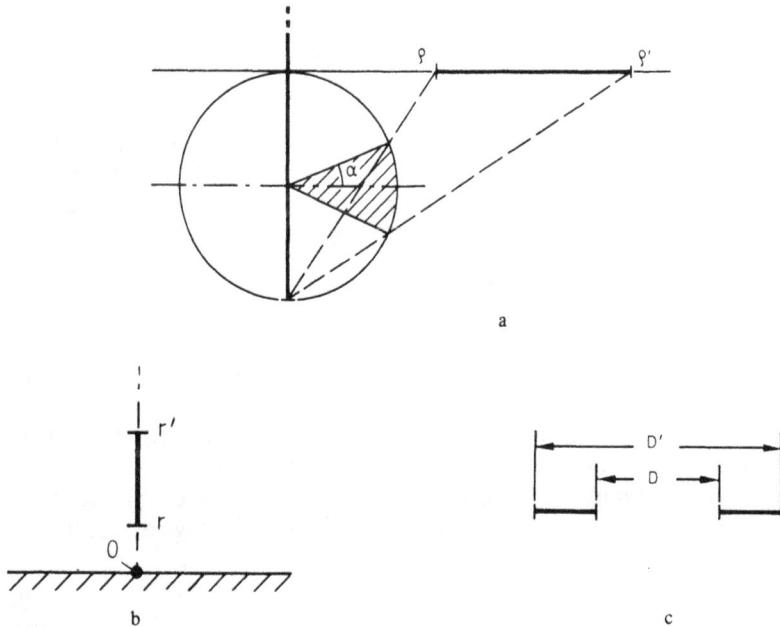

Figure 15

If the conical system of Figure 15a is now rotated around the horizontal axis in the plane of the drawing by the angle $\pi/2$, then a projection onto the plane as in Figure 16a produces a circular section opposed to a conductive plane as in Figure 16b. The angle α does not change in the projection, so that (85) also applies here without change. This is also the case for the two circular sections obtained by reflection which are opposed as in Figure 16c, for which (79) therefore is also valid without change.

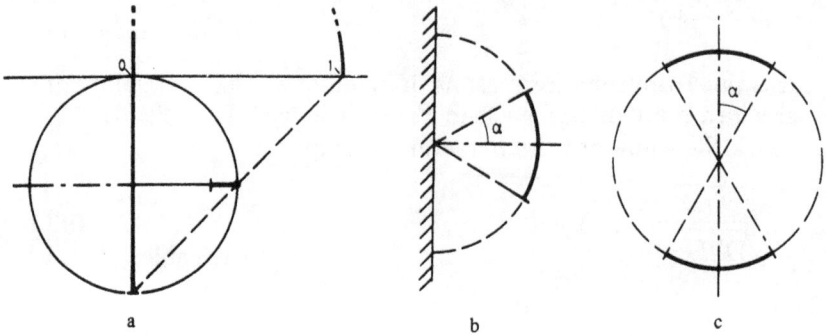

Figure 16

It still remains to supplement (73). The characteristic impedance of the conical strip opposed to the circular cone follows from the depiction of the strip in the circular cylinder from Figure 12 on the sphere. Again, (72) is valid. If these values are now inserted into (87), then it follows that:

$$k_A = \frac{\tan \alpha_1 / 2}{\tan \alpha_2 / 2}, \quad \lambda = \frac{1}{2}. \tag{94}$$

$$\text{(in Table II, No. 4)}$$

In such a conical stripline in a coaxial circular cone, Figure 12, the assumption of values $\alpha_2 \geqslant \pi/2$ is naturally also permitted, as long as $\alpha_1 > \alpha_2$. Specifically, for $\alpha_2 = \pi/2$, i.e., for a circular cone perpendicular to the edge of a conductive plane as in Figure 17, it follows that:

$$k_A = \tan \alpha_1 / 2, \quad \lambda = \frac{1}{2}. \tag{95}$$

$$\text{(in Table II, No. 7)}$$

These conical transmission lines can naturally also be projected onto the plane in truly diverse ways.

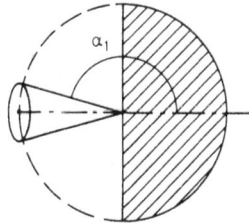

Figure 17

If the common axis of the cones of Figure 12 is perpendicular to the plane of the projection, no new systems are produced. However, if it is parallel to the plane of the projection, several new systems can be generated.

If $\alpha_1 > \pi/2$, then with parallel placement of the conical strip to the projection plane as in Figure 18a, there is formed a circular conductor opposed to a segment of a circle, Figure 18b. With placement of the conical strip parallel to the plane of the drawing as in Figure 18c, there is formed a circular conductor opposed to a strip oriented vertically to it, Figure 18d.

If $\alpha_1 < \pi/2$, a segment of a circle in a circular cylinder is first formed, and then a strip displaced from the center in a circular cylinder, in appropriate sequence as above.

However, the calculation for these systems will be done only later, where it is carried out with less restrictive conditions and also somewhat more simply.

However, for $\alpha_2 = \pi/2$ in the position of Figure 18c, the characteristic impedance of a circular cylinder opposed to the edge of a conductive plane as in Figure 19 can already be calculated easily and without restriction.

From Figure 18c, it follows for this case:

$$\rho_1 = \frac{r_1}{a} = \sqrt{\frac{r_1}{r_1'}} = \tan \frac{1}{2}\left(\alpha_1 - \frac{\pi}{2}\right). \tag{96}$$

Solution for $\alpha_1/2$ and insertion in (95) leads to:

$$k_A = \frac{\sqrt{r'/r + 1}}{\sqrt{r'/r - 1}}, \quad \lambda = \frac{1}{2}. \tag{97}$$

(in Table I, No. 17)

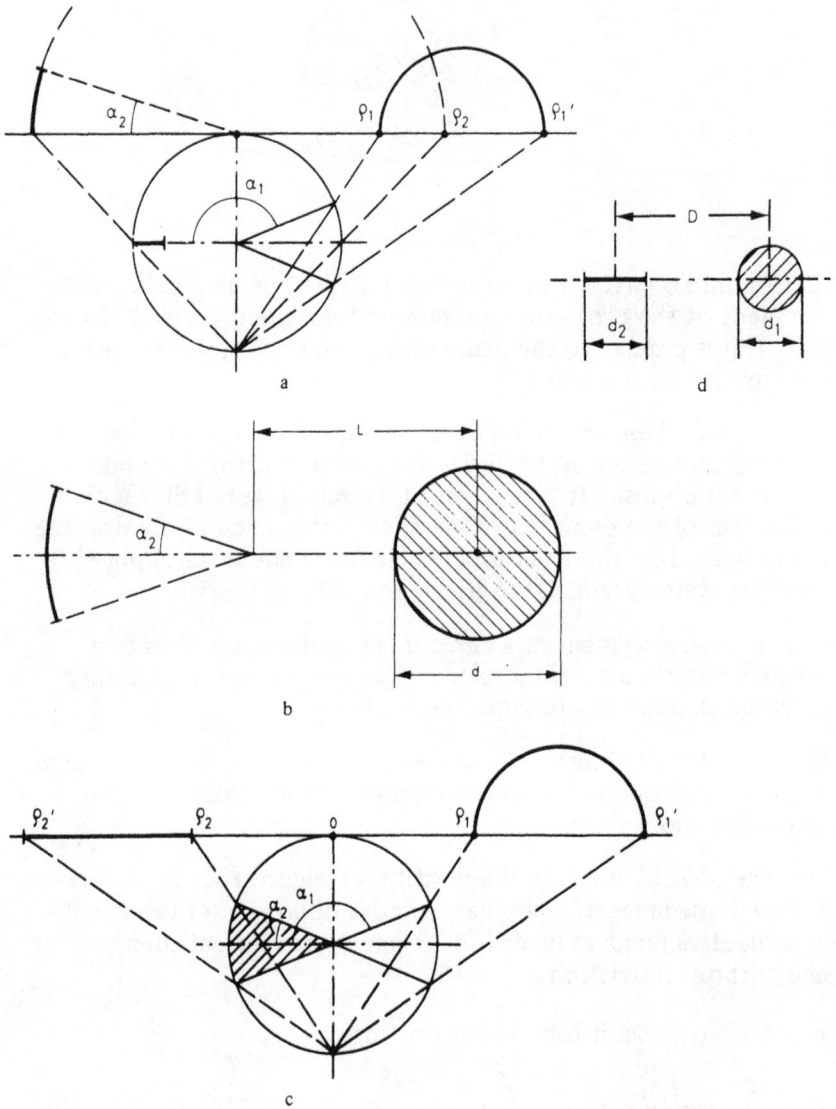

Figure 18

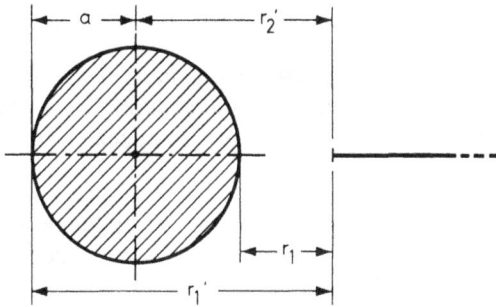

Figure 19

If the double conical strips from Figure 20a are projected onto the plane, a strip in the center of the opening of an otherwise conductive plane, Figure 20b, is obtained. However, the calculation does not need to be carried out in detail if it is noted that the same conductor results from an application of the complementarity rule of (14). In Figure 20c is sketched the previously calculated two-strip transmission line, on the bottom, and above it, the complementary line. For the latter, with the designations of Figure 20b and with the use of (14) on (93) (always after reverting to 82, naturally), the following applies:

$$k_A = \sqrt{D'/D}, \quad \lambda = 1. \tag{98}$$

(in Table I, No. 33)

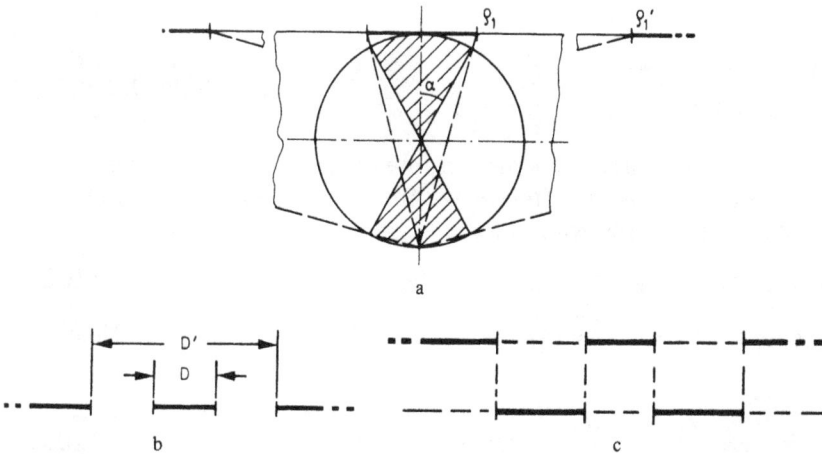

Figure 20

3. Conical strip and circular cone, anaxial

For the generation of the more general anaxial conical systems, it is apparently desirable to subject the available cross section of the cylindrical transmission lines to changes of scale, and then to portray them on a sphere.

If a strip perpendicular to a conductive plane, as in Figure 15b, is projected on a sphere as in Figure 21a, a conical strip is obtained which occupies an arbitrary position characterized by the angle γ in a plane perpendicular to the conductive base plane. It follows from the equations:

$$\rho = \tan \frac{1}{2}(\gamma - \alpha) = \frac{r}{a}, \quad \rho' = \tan \frac{1}{2}(\gamma + \alpha) = \frac{r'}{a}. \tag{99}$$

Division produces r'/r, and insertion of this value into (92) leads to:

$$k_A = \frac{\sqrt{\tan \frac{\gamma+\alpha}{2} \Big/ \tan \frac{\gamma-\alpha}{2}} + 1}{\sqrt{\tan \frac{\gamma+\alpha}{2} \Big/ \tan \frac{\gamma-\alpha}{2}} - 1}, \quad \lambda = \frac{1}{2} \tag{100}$$
(in Table II, No. 8)

For reasons of symmetry, the double conical strips of Figure 21b inclined to one another in a plane have twice the characteristic impedance, or:

$$k_A = \frac{\sqrt{\tan \frac{\gamma+\alpha}{2} \Big/ \tan \frac{\gamma-\alpha}{2}} + 1}{\sqrt{\tan \frac{\gamma+\alpha}{2} \Big/ \tan \frac{\gamma-\alpha}{2}} - 1}, \quad \lambda = 1 \tag{101}$$
(in Table II, No. 9)

The complementary transmission line consists of two coplanar, co-axial conical strips of different widths, as in Figure 21c. The following relationships apply to the angles:

$$\alpha_1 + \alpha_2 = \pi - 2\alpha, \quad \alpha_2 - \alpha_1 = \pi - 2\gamma \tag{102}$$

Solving for $(\gamma + \alpha)/2$ and $(\gamma - \alpha)/2$ and insertion into (101) leads to:

$$k_A = \cot \frac{\alpha_1}{2} \cot \frac{\alpha_2}{2}, \quad \lambda = 1 \tag{103}$$
(in Table II, No. 18)

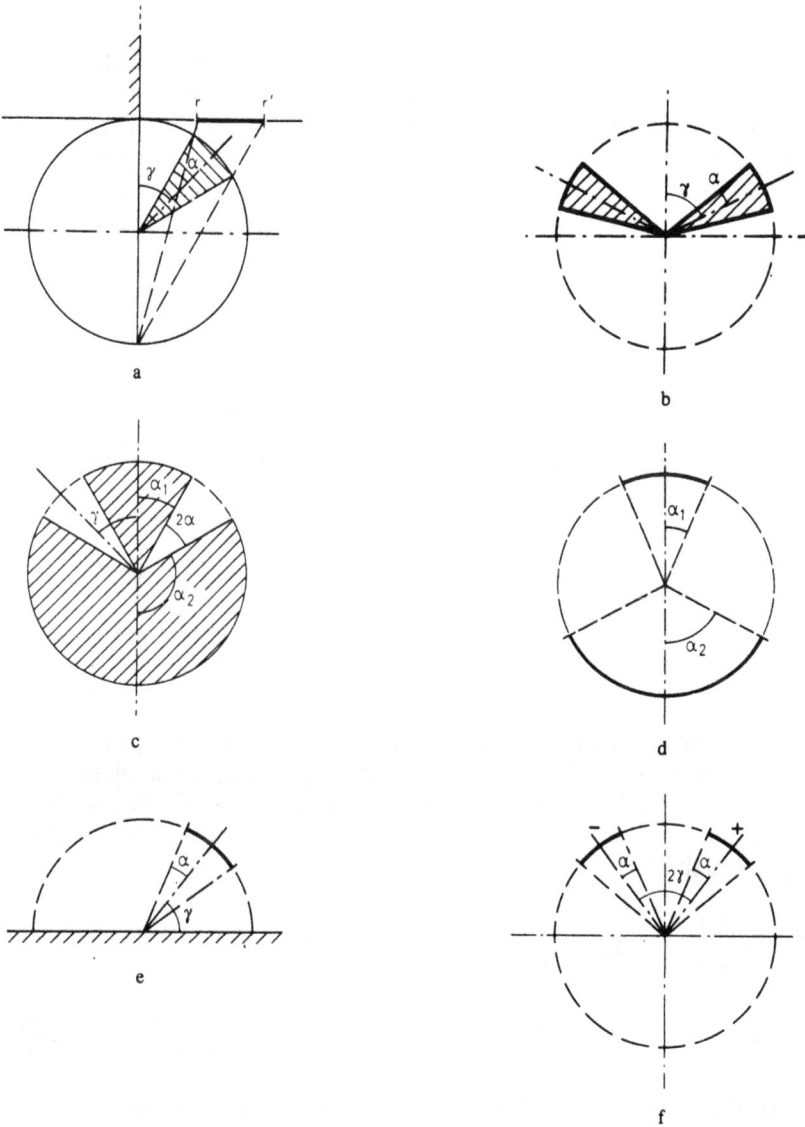

Figure 21

These conical transmission lines are again projected on a plane as a first step. If the sphere in Figure 21a is rotated around the horizontal axis in the plane of the drawing by the angle $\pi/2$, a segment of a circle opposed to a conductive plane as in Figure 21e is obtained by projection. The angles do not change in the projection, so that (100) is valid here without change. The same also results for the two identical segments of circles from Figure 21f, for which (101) is therefore valid, and for the opposite non-identical segments of circles from 21d, for which (103) is valid.

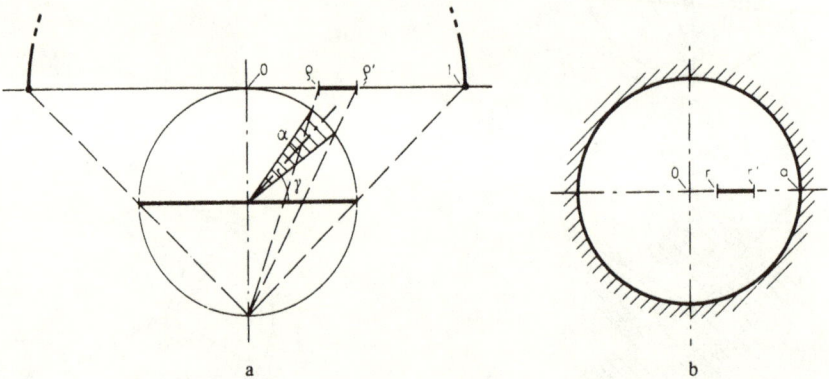

Figure 22

If the sphere in Figure 21a is rotated around the axis perpendicular to the plane of the drawing by the angle $\pi/2$, the projection in the position of Figure 22a produces a circular conductor with an internal strip displaced from the center, Figure 22b. It follows from this, that:

$$\rho = \tan \frac{1}{2}\left(\frac{\pi}{2} - \gamma - \alpha\right) = \frac{r}{a},$$

$$\rho' = \tan \frac{1}{2}\left(\frac{\pi}{2} - \gamma + \alpha\right) = \frac{r'}{a}. \tag{104}$$

The solution for $(\gamma + \alpha)/2$ and $(\gamma - \alpha)/2$ and insertion into (100) leads to the parameters:

$$k_A = \frac{\sqrt{(a-r)(a+r')/(a+r)(a-r')} + 1}{\sqrt{(a-r)(a+r')/(a+r)(a-r')} - 1}, \quad \lambda = \frac{1}{2}. \tag{105}$$

(in Table I, No. 14)

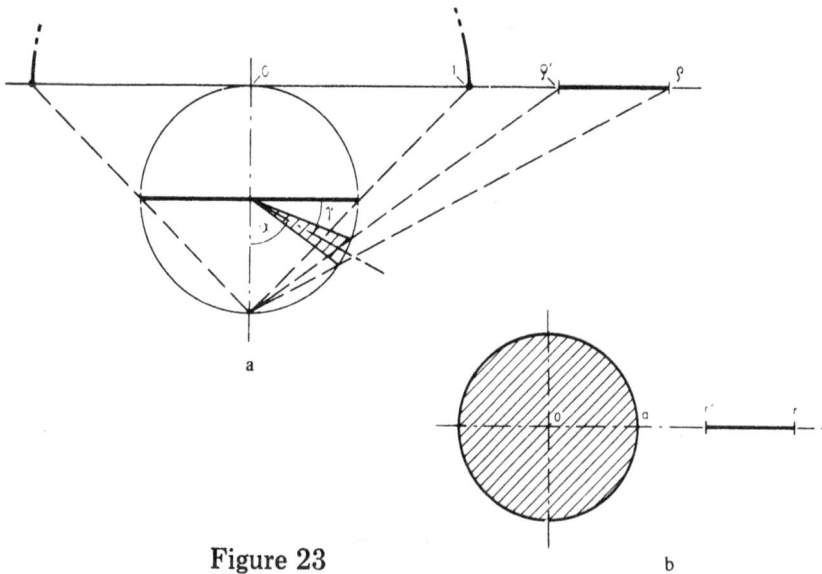

Figure 23

In the configuration of Figure 23a, projection produces a circular conductor with a strip oriented vertically to it, Figure 23b. From this, it follows that

$$\rho' = \tan \frac{1}{2}\left(\frac{\pi}{2} + \gamma - \alpha\right) = \frac{r'}{a} ,$$

$$\rho = \tan \frac{1}{2}\left(\frac{\pi}{2} + \gamma + \alpha\right) = \frac{r}{a} . \tag{106}$$

Solution for $(\gamma + \alpha)/2$ and $(\gamma - \alpha)/2$ and insertion into (100) produces the parameters:

$$k_A = \frac{\sqrt{(r-a)(r'+a)/(r+a)(r'-a)} + 1}{\sqrt{(r-a)(r'+a)/(r+a)(r'-a)} - 1} , \quad \lambda = \frac{1}{2} . \tag{107}$$

(in Table I, No. 13)

If $r \to \infty$, the case is obtained of a circular conductor opposed to the edge of a conductive plane, and it is easily verified that the known parameters of (97) are again produced with the designations $r'_1 = r'_2 + a$; $r_1 = r'_2 - a$ from Figure 19.

The system of Figure 19 can now be portrayed again onto a sphere as in Figure 24, whereby there is obtained the circular cone fixed in position by the angle γ, opposed to the edge of a conductive plane. From this it follows that:

$$\rho = \tan \frac{\gamma - \alpha}{2} = \frac{r}{a}, \quad \rho' = \tan \frac{\gamma + \alpha}{2} = \frac{r'}{a}. \tag{108}$$

Division produces r'/r and insertion into (97) leads to:

$$k_A = \frac{\sqrt{\tan \dfrac{\gamma + \alpha}{2} / \tan \dfrac{\gamma - \alpha}{2} + 1}}{\sqrt{\tan \dfrac{\gamma + \alpha}{2} / \tan \dfrac{\gamma - \alpha}{2} - 1}}, \quad \lambda = \frac{1}{2}. \tag{109}$$

(in Table II, No. 10)

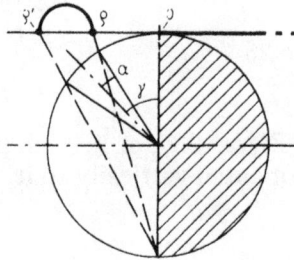

Figure 24

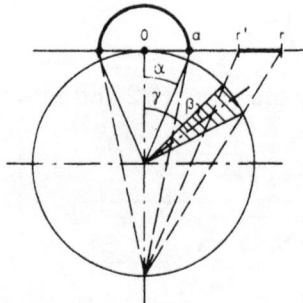

Figure 25

It is recognized that the same expressions appear frequently, e.g., equations (109) and (100), and (97) and (92) are identical with the chosen designations. The systems in Figure 24 and Figure 21a as well as those in Figure 19 and Figure 15b, however, are different from one another.

The general expression for the impedance of a circular cone opposed to a strip cone lying in the plane formed by the two axes is obtained by depicting the flat system of Figure 23b on a sphere as in Figure 25. It is found here, if n signifies a normalization parameter:

$$\frac{a}{n} = \tan \frac{\alpha}{2}, \quad \frac{r'}{n} = \tan \frac{\gamma-\beta}{2}, \quad \frac{r}{n} = \tan \frac{\gamma+\beta}{2}. \tag{110}$$

By insertion in (107), the normalization again drops out, and it follows that:

$$k_A = \frac{\sqrt{\dfrac{[\tan (\gamma+\beta)/2 - \tan \alpha/2][\tan (\gamma-\beta)/2 + \tan \alpha/2]}{[\tan (\gamma+\beta)/2 + \tan \alpha/2][\tan (\gamma-\beta)/2 - \tan \alpha/2]}} + 1}{\sqrt{\dfrac{[\tan (\gamma+\beta)/2 - \tan \alpha/2][\tan (\gamma-\beta)/2 + \tan \alpha/2]}{[\tan (\gamma+\beta)/2 + \tan \alpha/2][\tan (\gamma-\beta)/2 - \tan \alpha/2]}} - 1},$$

$$\lambda = \frac{1}{2}. \tag{111}$$

(in Table II, No. 11)

The formula has already become truly comprehensive with these three determining parameters α, β, and γ.

By rotation of the horizontal axis of the sphere in Figure 24 lying in the plane of the drawing by the angle $\pi/2$, from the projection in accordance with Figure 26a can be obtained the cylindrical transmission line of a full circle opposed to an open semicircle, Figure 26b.

There is no change in the angles α, γ from this projection, so that (109) is valid for this without change. According to Figure 26b, care must only be taken in the construction that the following condition is met:

$$r \cdot r' = a^2. \tag{112}$$

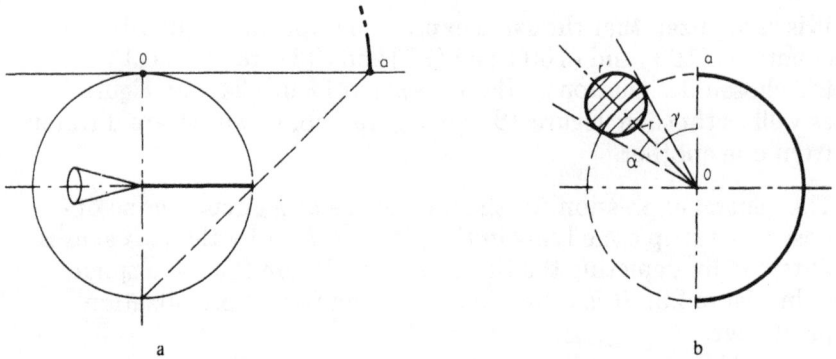

Figure 26

The same production can also be carried out with the conical transmission line of Figure 25. Analogous to Figure 26b, the system of any arbitrary segment of a circle and a circle obeying only the condition (112) is produced in any orientation, with Equation (111) applying to it.

In the portrayal of the plane on the sphere, the conditions have been selected so that conical strips are formed in great circle surfaces. However, projections where this is not the case also seem to be of interest. As an example, let us start with the two circular segments in Figure 16c or Figure 21d, select the radius of the circle smaller than a, and then carry out a projection as in Figure 27a. The slotted circular cone line of Figure 27b is formed. Since the angles in plane parallel to the plane of projection do not change, Figures 16c and 21d can also be considered at the same time as (planar) cross sections through conical lines with the equations (79) and (103) for the characteristic impedances then also unchanged for this case. This interpretation is then apparently also correct for other examples of circular segments on a circle.

Above all, it is astonishing that nothing more depends, after all, on the aperture angle of the circular cone. This becomes particularly obvious when one imagines himself tied to a cross section, such as in Figure 16c, for example, and the feed point of the associated conical line is then chosen outward from the center perpendicular to the plane of the drawing at a different distance from the plane of the cross section. Finally, if the distance is allowed to approach infinity, the cylindrical transmission line is again obtained in a continuous transition.

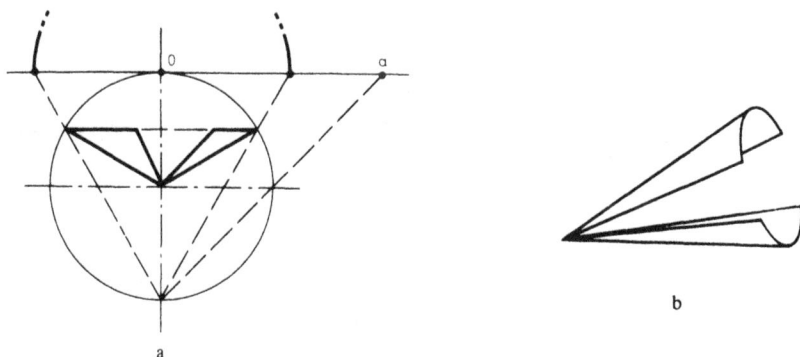

Figure 27

4. The Use of the Transformation $w = z^n$

One of the simplest conformal mappings is $w = z^n$. For $n \neq \pm 1$, it becomes a non-linear transformation and then in particular no longer belongs in the group of circular relationships. Although the deformation of transmission line cross sections caused by such a transformation in general leads to poorly comprehensible new cross sections, it can nevertheless be used to advantage for some striplines. That is to say, since circles around the origin and straight lines through the origin retain their nature in this transformation, in the case of transmission line cross sections which consist only of the circles and lines mentioned or sections of them, or which are bounded by such sections, particularly closely related, easily comprehensible transformed systems are formed. In the use of this transformation, care must simply be taken that a volume of space charged with field energy never disappears from the first layer of the z-plane.

To start with, let the system of the circular segment opposed to a conductive plane from Figure 21e be subjected to this transformation. For $n > 1$, there are formed circular segments opposed to an acute angle or opposed to the edge of a conductive plane (Figure 28a and b), and for $n < 1$, circular segments in a reentrant angle (Figure 28c). This is because:

$$z = r_1 \, e^{i\alpha_1}, \quad w = z^n = r_1^n \, e^{in\alpha_1} = r_2 \, e^{i\alpha_2} \qquad (113a)$$

$$r_2 = r_1^n, \quad \alpha_2 = n\alpha_1, \quad \gamma_2 = n\gamma_1. \qquad (113b)$$

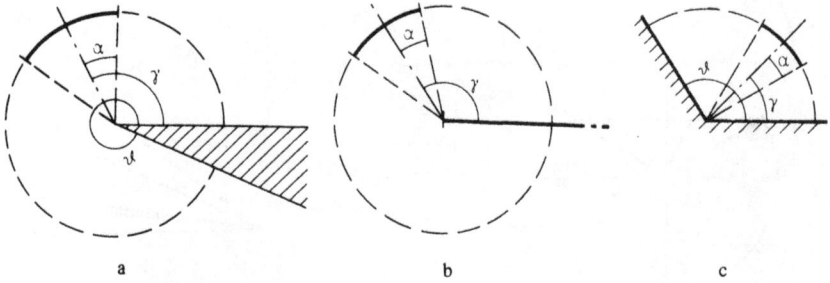

Figure 28

Insertion of the last equation into (100) produces:

$$k_A = \frac{\sqrt{\tan \dfrac{\gamma_2 + \alpha_2}{2n} \Big/ \tan \dfrac{\gamma_2 - \alpha_2}{2n} + 1}}{\sqrt{\tan \dfrac{\gamma_2 + \alpha_2}{2n} \Big/ \tan \dfrac{\gamma_2 - \alpha_2}{2n} - 1}}, \quad \lambda = \frac{1}{2}. \tag{114}$$

The value n can now be expressed by the angle ϑ, if it is kept in mind that the following must be true

$$\vartheta_2 = n\pi. \tag{115}$$

In this way are obtained the parameters:

$$k_A = \frac{\sqrt{\tan \dfrac{\pi}{2} \dfrac{\gamma + \alpha}{\vartheta} \Big/ \tan \dfrac{\pi}{2} \dfrac{\gamma - \alpha}{\vartheta} + 1}}{\sqrt{\tan \dfrac{\pi}{2} \dfrac{\gamma + \alpha}{\vartheta} \Big/ \tan \dfrac{\pi}{2} \dfrac{\gamma - \alpha}{\vartheta} - 1}}, \quad \lambda = \frac{1}{2}. \tag{116}$$

(in Table I, No. 19)

For the special case $\vartheta = 2\pi$ as in Figure 28b, it follows that:

$$k_A = \frac{\sqrt{\tan \dfrac{\gamma + \alpha}{4} \Big/ \tan \dfrac{\gamma - \alpha}{4} + 1}}{\sqrt{\tan \dfrac{\gamma + \alpha}{4} \Big/ \tan \dfrac{\gamma - \alpha}{4} - 1}}, \quad \lambda = \frac{1}{2}. \tag{117}$$

(in Table I, No. 20)

It also appears to be of interest that for $\gamma = \pi$, $\alpha = \pi/2$, the characteristic impedance is always exactly $Z = \eta/4$. If the angle ϑ of the roontrant angle of 28c is so chosen that a whole multiple again gives π, or:

$$N \cdot \vartheta = \pi, \quad N \geqslant 2, \tag{118}$$

then as is known from the reflection principle, systems without separating walls can be generated as in Figure 29a, since then because of (116), the parameters have:

$$k_A = \frac{\sqrt{\tan \frac{N}{2}(\gamma+\alpha)/\tan \frac{N}{2}(\gamma-\alpha) + 1}}{\sqrt{\tan \frac{N}{2}(\gamma+\alpha)/\tan \frac{N}{2}(\gamma-\alpha) - 1}},$$

$$\lambda = \frac{2}{N}, \quad N \geqslant 2, \tag{119}$$

(in Table I, No. 22)

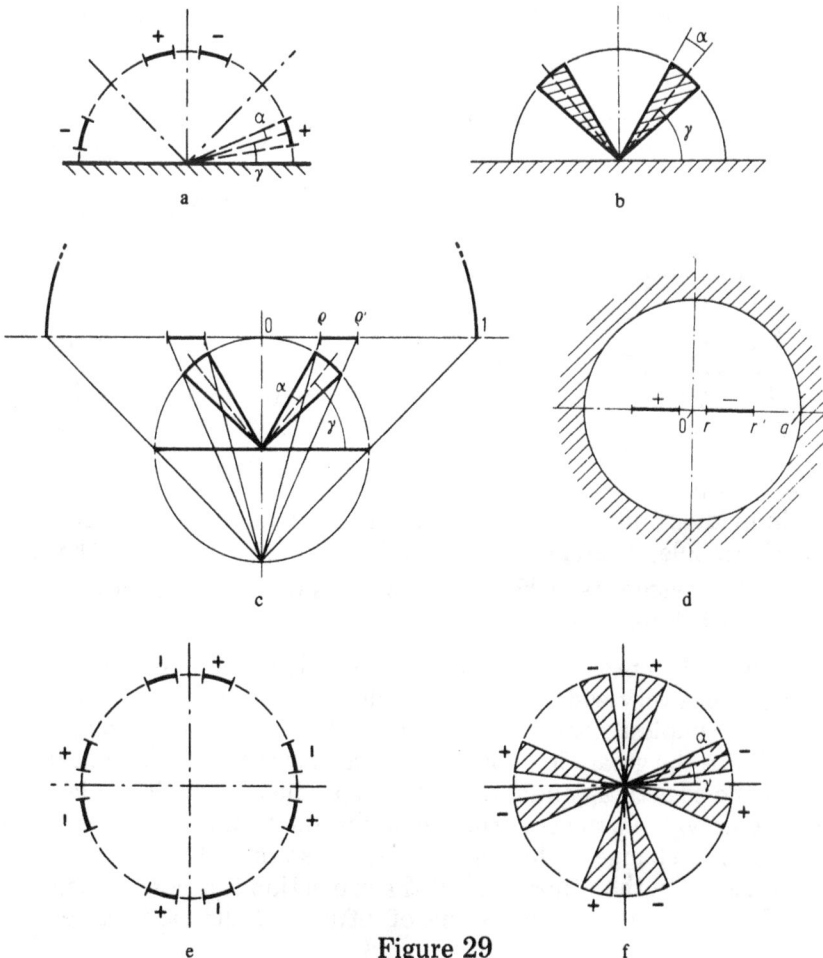

Figure 29

N here is the number of strips above the conductive plane. Specifically for N = 2, another very useful system will later be produced by a further transformation. For that reason, its parameters might also be indicated here explicitly:

$$k_A = \frac{\sqrt{\tan(\gamma+\alpha)/\tan(\gamma-\alpha)} + 1}{\sqrt{\tan(\gamma+\alpha)/\tan(\gamma-\alpha)} - 1} , \lambda = 1. \qquad (120a)$$

(in Table I, No. 22)

If Figure 29a is converted into a conical strip system, then Figure 29b is obtained (drawn with only two strips). If this line is then projected into the orientation of Figure 29c onto a plane, a line with the cross section of Figure 29d is obtained. The projection equations follow:

$$\rho = \frac{r}{a} = \tan\frac{1}{2}\left(\frac{\pi}{2} - \gamma - \alpha\right) , \ \rho' = \frac{r'}{a} = \tan\frac{1}{2}\left(\frac{\pi}{2} - \gamma + \alpha\right), \qquad (120b)$$

Solution for $(\gamma + \alpha)$ and $(\gamma - \alpha)$ and insertion into (120a) leads to the parameters:

$$k_A = \frac{\sqrt{\dfrac{r'}{r}\dfrac{a^2-r^2}{a^2-r'^2}} + 1}{\sqrt{\dfrac{r'}{r}\dfrac{a^2-r^2}{a^2-r'^2}} - 1} , \lambda = 1. \qquad (120c)$$

(in Table I, No. 46)

By reflection on the conductive plane of Figure 29a, a conductor with individual conductive segments can be produced on a cylinder as in Figure 29e, wherein (119) is valid for k_A with $\lambda = 1/N$. The conical strip system as in Figure 29f also has the same value of characteristic impedance.

If it is desired to project the flat system of Figure 28b onto a sphere, then a very wide variety of conical systems can be generated. The simplest system occurs if the circular segment is transformed into the great circle of a sphere as in Figure 30. A conical strip at the edge of a conductive plane is produced, which can be swiveled at will in a plane perpendicular to the hemiplane. The angles do not change in the transformation, so that (117) also applies here. On the other hand, if the projection is accomplished as in Figure 30b, two conical strips of different widths are formed,

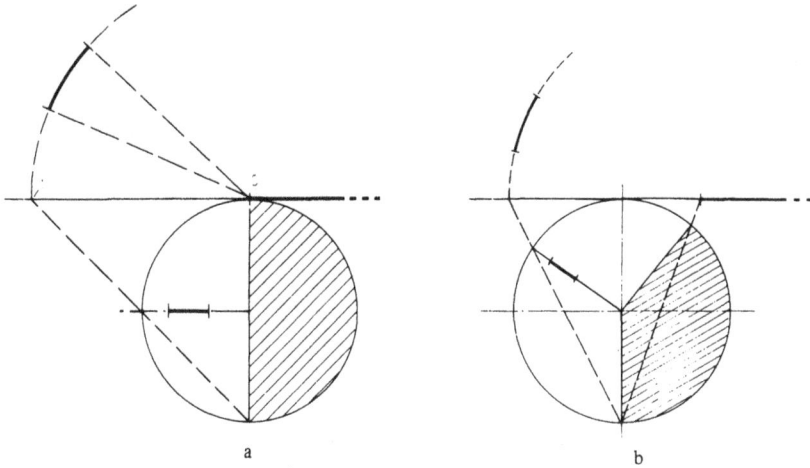

Figure 30

capable of being swiveled arbitrarily in planes perpendicular to one another, which will not be discussed in further detail here.

If the cone system formed from Figure 30a is rotated around the horizontal axis in the plane of the drawing by the angle $\pi/2$, a projection as in Figure 31a produces a strip lying on the axis of symmetry, Figure 31. From this it can be deduced that:

$$\rho = \tan \frac{1}{2}\left(\gamma - \alpha - \frac{\pi}{2}\right) = \frac{r}{a}, \rho' = \tan \frac{1}{2}\left(\gamma + \alpha - \frac{\pi}{2}\right) = \frac{r'}{a}.$$

This can be solved for $(\gamma + \alpha)/4$ and $(\gamma - \alpha)/4$ and substituted in (117). The formula becomes so extensive from this, that it will not be reproduced here.

If the axis of the sphere perpendicular to the plane of the drawing in Figure 31a is rotated by an angle $\pi/2$ in the positive direction, there is produced a projection as in 32a of two strips oriented perpendicular to one another, as in 32b. From this it follows that:

$$\rho = \tan \frac{1}{2}\left(\gamma - \alpha\right) = \frac{r}{a}, \quad \rho' = \tan \frac{1}{2}\left(\gamma + \alpha\right) = \frac{r'}{a}. \qquad (122)$$

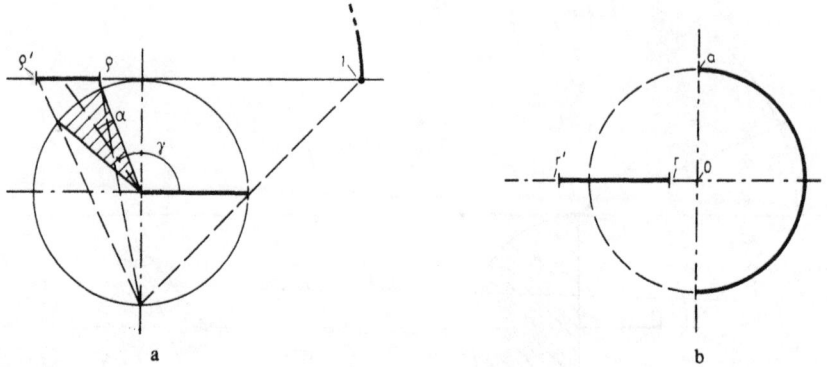

Figure 31

Solution for $(\gamma + \alpha)/4$ and $(\gamma - \alpha)/4$ after use of the addition theorem:

$$\tan \alpha/2 = \tan \alpha/(1 + \sqrt{1 + \tan^2 \alpha})$$ (123)

after substitution in (117) leads to:

$$k_A = \frac{\sqrt{\dfrac{r'}{r} \dfrac{1 + \sqrt{1 + r^2/a^2}}{1 + \sqrt{1 + r'^2/a^2}} + 1}}{\sqrt{\dfrac{r'}{r} \dfrac{1 + \sqrt{1 + r^2/a^2}}{1 + \sqrt{1 + r'^2/a^2}} - 1}}, \quad \lambda = \frac{1}{2}.$$ (124)
(in Table I, No. 24)

If $r' \to \infty$, we obtain the strip transverse to the edge of a conductive plane as in Figure 32c, with the characteristic impedance:

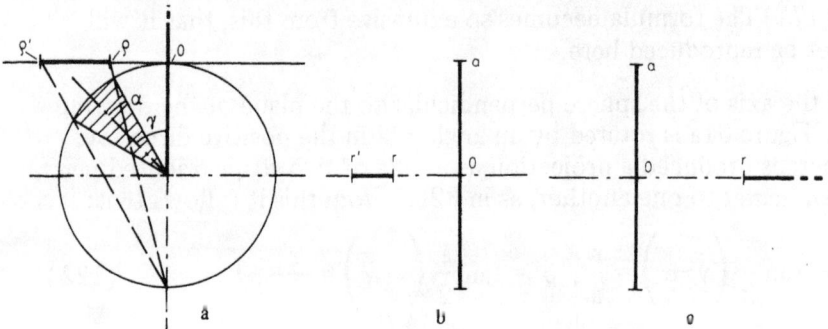

Figure 32

$$k_A = \cfrac{\sqrt{\dfrac{a}{r}\left(1 + \sqrt{1 + r^2/a^2}\right) + 1}}{\sqrt{\dfrac{a}{r}\left(1 + \sqrt{1 + r^2/a^2}\right) - 1}}, \quad \lambda = \frac{1}{2}. \qquad \begin{array}{l}(125)\\ \text{(in Table I, No. 25)}\end{array}$$

If the axis of the sphere perpendicular to the plane of the drawing of Figure 31a is rotated in the negative direction by the angle $\pi/2$, a projection as in 33a produces a strip transverse to the opening in a conductive plane, Figure 33b. From this it follows that:

$$\rho = \tan \frac{1}{2}(\pi - \gamma - \alpha) = \frac{r}{a}, \quad \rho' = \tan \frac{1}{2}(\pi - \gamma + \alpha) = \frac{r'}{a}. \qquad (126)$$

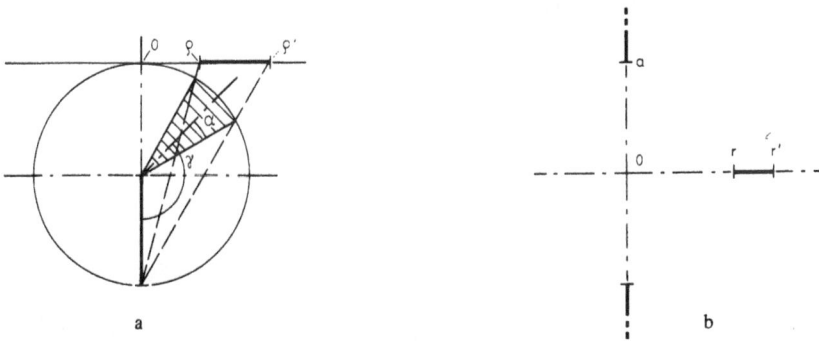

Figure 33

Solution for $(\gamma + \alpha)/4$ and $(\gamma - \alpha)/4$, after application of the addition theorem:

$$\cot \alpha/2 = \cot \alpha + \sqrt{1 + \cot^2 \alpha} \qquad (127)$$

and substitution in (117) leads to:

$$k_A = \cfrac{\sqrt{\dfrac{r' + \sqrt{a^2 + r'^2}}{r + \sqrt{a^2 + r^2}} + 1}}{\sqrt{\dfrac{r' + \sqrt{a^2 + r'^2}}{r + \sqrt{a^2 + r^2}} - 1}}, \quad \lambda = \frac{1}{2}. \qquad \begin{array}{l}(128)\\ \text{(in Table I, No. 26)}\end{array}$$

If $a \to 0$ here as a test (or in (124) $a \to \infty$), (92) is again obtained.

For expediency, let us assume that the system of the strip perpendicular to a plane of Figure 15b is subjected to the transformation $w = z^n$. The exponent n may again assume the values $0 < n < 2$, where a strip opposed to an acute angle as in Figure 34a is produced for $1 < n < 2$, a strip opposed to the edge of a conductive plane as in 34b for $n = 2$, and a strip on the angle bisector of a reentrant angle as in Figure 34c for $0 < n < 1$. The transformation of the section lying on the radial rays must be kept especially in view here.

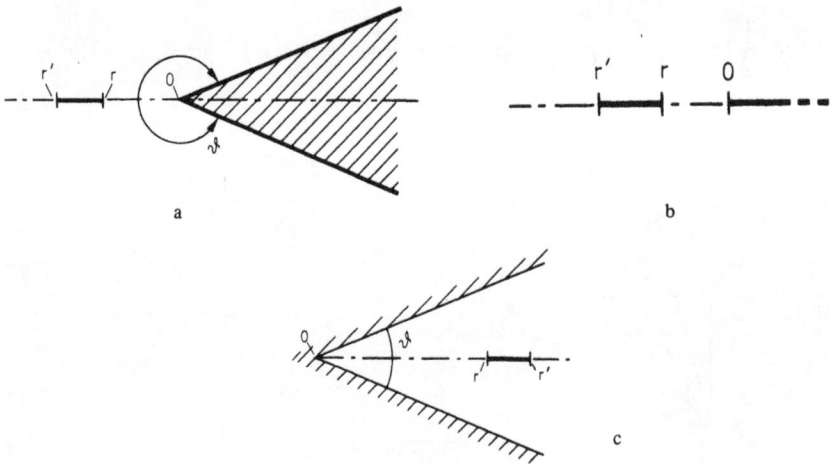

Figure 34

From (113), it is:

$$r_2 = r_1{}^n , \quad r_2' = r_1'{}^n .$$

$$(129)$$

Solution for r_1 and r_1' and substitution in (92) produces the parameters:

$$k_A = \frac{(r'/r)^{1/2n} + 1}{(r'/r)^{1/2n} - 1} , \quad n = \frac{\vartheta}{\pi} , \quad \lambda = \frac{1}{2} .$$

$$(130)$$
(in Table I, No. 27)

With observance of the condition (118), N strips of Figure 34c can again be united in a hemiplane without separating walls, as in Figure 35a.

Figure 35

For the parameters of the overall system, the following is then valid:

$$k_A = \frac{(r'/r)^{N/2} + 1}{(r'/r)^{N/2} - 1} \;,\quad \lambda = \frac{2}{N}\;,\quad N \geqslant 2\;.$$

$$(131)$$
$$\text{(in Table I, No. 29)}$$

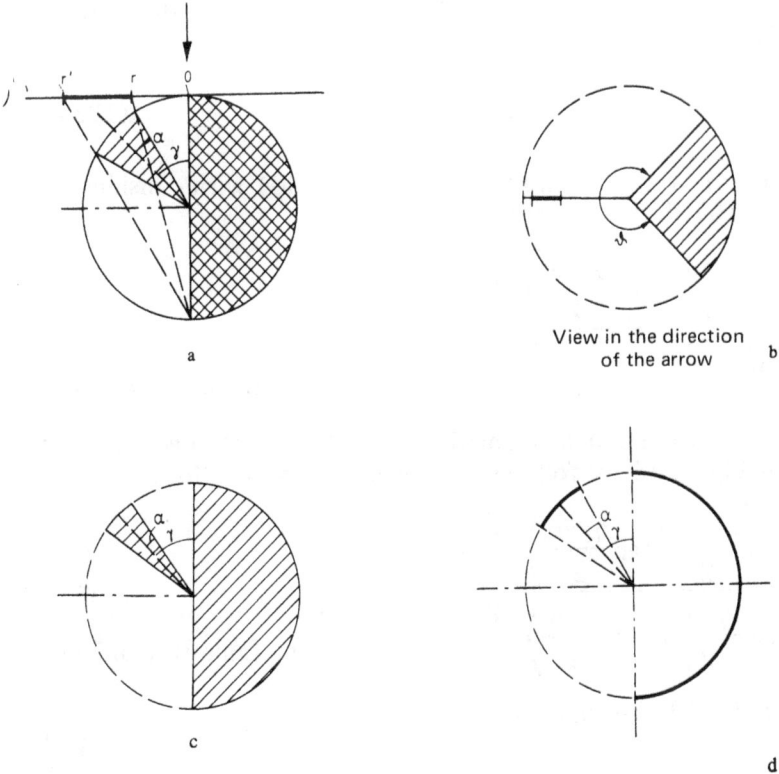

View in the direction
of the arrow

Figure 36

The circular conductor formed by reflection from radially oriented strips, as in Figure 35b for four strips (N = 2), then has half the value as its characteristic impedance, i.e., in comparison with (131), $\lambda = 1/N$.

For the strip opposed to the edge of a conductive plane in Figure 34b, for N = 2 in particular, there results from (130):

$$k_A = \frac{\sqrt[4]{r'/r} + 1}{\sqrt[4]{r'/r} - 1} , \quad \lambda = \frac{1}{2} . \tag{132}$$
(in Table I, No. 28)

This line had played a central role in [1], in addition to the double conical transmission line, since it retained its character in complementation.

If the cylindrical transmission line of Figure 34a is portrayed on a sphere, as in 36a, a conical strip opposed to a wedge is produced, Figure 36b. From this it follows that:

$$\rho = \tan \frac{1}{2} (\gamma - \alpha) = \frac{r}{a} , \quad \rho' = \tan \frac{1}{2} (\gamma + \alpha) = \frac{r'}{a} . \tag{133}$$

By forming the ratio r'/r and substitution in (130), with consideration of (115), there is obtained:

$$k_A = \frac{\left(\tan \frac{\gamma + \alpha}{2} / \tan \frac{\gamma - \alpha}{2} \right)^{\pi/2\vartheta} + 1}{\left(\tan \frac{\gamma + \alpha}{2} / \tan \frac{\gamma - \alpha}{2} \right)^{\pi/2\vartheta} - 1} , \quad \lambda = \frac{1}{2} . \tag{134}$$
(in Table II, No. 15)

For $\vartheta = 2\pi$, for the coplanar, anaxial, conical strips of unequal widths of Figure 36c, it follows specifically from this, that:

$$k_A = \frac{\left(\tan \frac{\gamma + \alpha}{2} / \tan \frac{\gamma - \alpha}{2} \right)^{1/4} + 1}{\left(\tan \frac{\gamma + \alpha}{2} / \tan \frac{\gamma - \alpha}{2} \right)^{1/4} - 1} , \quad \lambda = \frac{1}{2} . \tag{135}$$
(in Table II, No. 16)

For $\gamma = \pi/2$, with the transformation:

$$\sqrt{\frac{\left(\tan \dfrac{\pi}{4} + \dfrac{\alpha}{2}\right)}{\left(\tan \dfrac{\pi}{4} - \dfrac{\alpha}{2}\right)}} = \sqrt{\frac{\left(\tan \dfrac{\pi}{4} + \dfrac{\alpha}{2}\right)}{\left(\cot \dfrac{\pi}{4} + \dfrac{\alpha}{2}\right)}} = \tan\left(\dfrac{\pi}{4} + \dfrac{\alpha}{2}\right) \qquad (136)$$

the parameters then follow:

$$k_A = \sqrt{\frac{\tan\left(\dfrac{\pi}{4} + \dfrac{\alpha}{2}\right) + 1}{\tan\left(\dfrac{\pi}{4} + \dfrac{\alpha}{2}\right) - 1}} \quad , \quad \lambda = \dfrac{1}{2}. \qquad \begin{array}{l}(137)\\ \text{(in Table II, No. 17)}\end{array}$$

This is obviously a new formula for a special case of the system in Figure 21c with (103).

In the projection onto a parallel tangential plane, the angles do not change, so that the same equations apply also to the corresponding circular segments of Figure 36d.

By projection of the conical strip transmission line of Figure 36c into the position of Figure 37a on the plane, the characteristic impedance of two coplanar parallel strips of unlike widths can easily be obtained, as in Figure 37b.

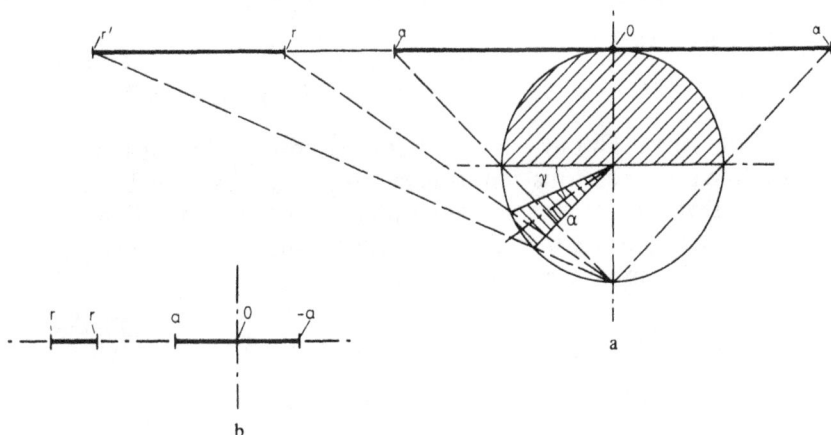

Figure 37

With the transformation equations:

$$\rho = \tan \frac{1}{2}\left(\frac{\pi}{2} + \gamma - \alpha\right) = \frac{r}{a},$$

$$\rho' = \tan \frac{1}{2}\left(\frac{\pi}{2} + \gamma + \alpha\right) = \frac{r'}{a} \tag{138}$$

there is initially produced:

$$\tan \frac{\gamma - \alpha}{2} = \tan\left(-\frac{\pi}{4} + \text{arc tan } \frac{r}{a}\right),$$

$$\tan \frac{\gamma + \alpha}{2} = \tan\left(-\frac{\pi}{4} + \text{arc tan } \frac{r'}{a}\right). \tag{139}$$

Substitution in (135) and resolution of the trigonometric functions provides the parameters:

$$k_A = \frac{\left(\dfrac{r'-a}{r'+a} \dfrac{r+a}{r-a}\right)^{1/4} + 1}{\left(\dfrac{r'-a}{r'+a} \dfrac{r+a}{r-a}\right)^{1/4} - 1}, \quad \lambda = \frac{1}{2}. \tag{140}$$
$$\text{(in Table I, No. 31)}$$

The complementary line is a coplanar, three-strip line with two slots of unequal widths. With (14), it follows for it:

$$k_A = \left(\frac{r'-a}{r'+a} \frac{r+a}{r-a}\right)^{1/4}, \quad \lambda = 2 \tag{141}$$
$$\text{(in Table I, No. 32)}$$

Figures 38a to e is an outline with a few examples of what transmission line cross sections can be generated easily by projection of the strip wedge systems of Figure 36a, b onto a plane. However, there is no limit yet in sight for the number of line cross sections which can be produced here.

5. *The Use of the Portrayal of a Hemiplane in a Strip*

By the well-known transformation [9] in the form:

$$w = ih - \frac{h}{\pi} \ln \frac{z + ia}{z - ia} \tag{142}$$

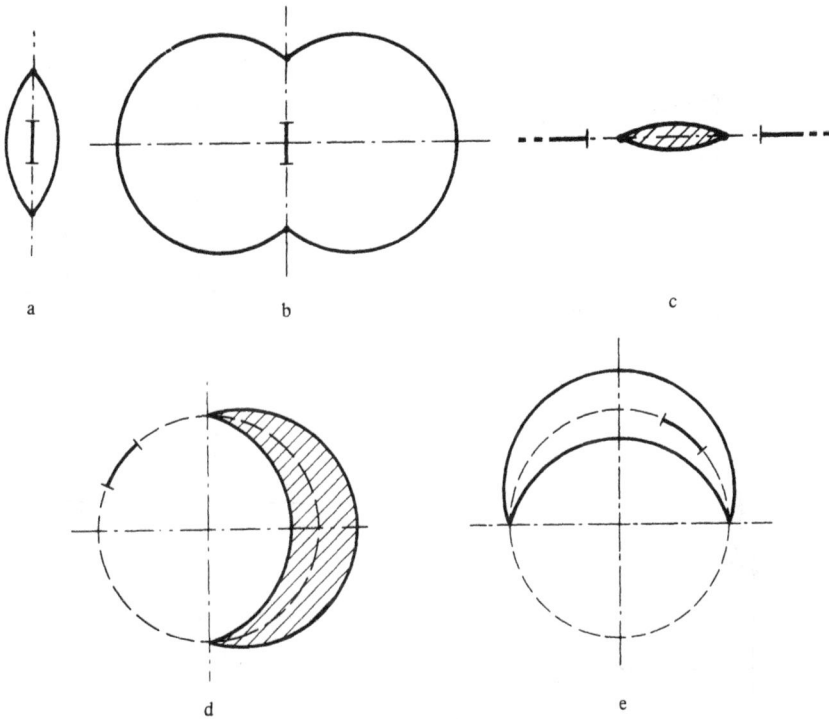

Figure 38

it has been possible, as in Figure 39a, b, to portray the right z-hemiplane as a strip of height h. It is seen immediately that the point z = + 1a becomes w = ih − ∞, and the point z = −ia becomes w = ih + ∞. If one now considers a semicircle around the origin of the z-plane which passes through the point z = ±ia, the substitution of:

$$z = ae^{i\beta}, \quad -\frac{\pi}{2} \leqslant \beta \leqslant \frac{\pi}{2} \tag{143}$$

in (142) first gives:

$$w = ih - \frac{h}{\pi} \ln \frac{e^{i\beta} + e^{i\pi/2}}{e^{i\beta} - e^{i\pi/2}} = ih - \frac{h}{\pi} \ln \coth i\left(\frac{\beta}{2} - \frac{\pi}{4}\right), \tag{144}$$

and finally with

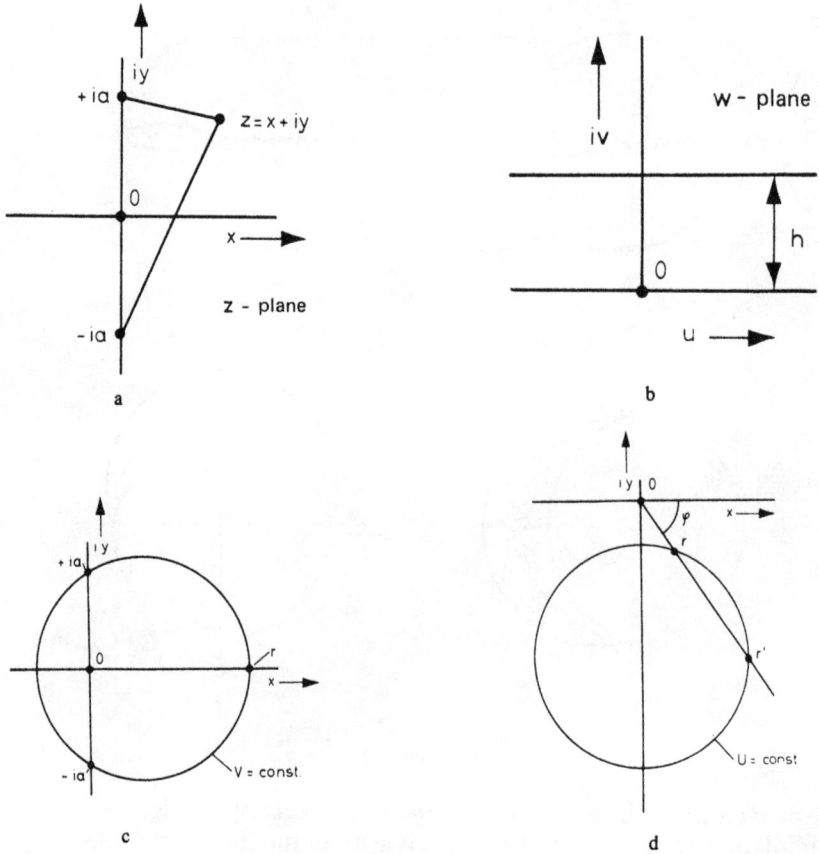

a

b

c

d

Figure 39

$$\coth ix = -i \cot x \qquad (145)$$

the center line of the strip is obtained:

$$w = u + iv = i\,\frac{h}{2} - \frac{h}{\pi}\ln \cot\left(\frac{\pi}{4} - \frac{\beta}{2}\right). \qquad (146)$$

The real positive axis in the z-plane:

$$z = r, \quad 0 \leqslant r \leqslant \infty \qquad (147)$$

by substitution in (142) and with the use of equation:

$$\text{arc tan } z = \frac{1}{2i}\ln\frac{1+iz}{1-iz} \qquad (148)$$

is portrayed in the section of the imaginary axis of the w-plane
with the equation:

$$w = iv = ih \left(1 - \frac{2}{\pi} \text{ arc tan } \frac{a}{r}\right). \tag{149}$$

In order to recognize further that all circles through the points
$z = \pm ia$ become straight lines parallel to the real axis in the w-plane,
the transformation equation (142) should be studied in somewhat
more detail. First, only the characteristic portion is considered:

$$w = \ln \frac{z+ia}{z-ia} = \ln \frac{1 + i\dfrac{a}{z}}{1 - i\dfrac{a}{z}}. \tag{150}$$

From this with (148):

$$w = 2i \text{ arc tan } \frac{a}{z} = 2i \text{ arc cot } \frac{z}{a}. \tag{151}$$

Solution for z, with $w = U + iV$, produces:

$$z = a \cot \frac{w}{2i} = a \cot \frac{U+iV}{2i} = a \cot \left(\frac{V}{2} - i\frac{U}{2}\right). \tag{152}$$

With the addition theorem:

$$\cot (x + iy) = -\frac{\sin 2x - i \sinh 2y}{\cos 2x - \cosh 2y} \tag{153}$$

further transformation can be carried out to give:

$$z = a \frac{\sin V + i \sinh U}{\cosh U - \cos V}. \tag{154}$$

The two coordinates x, y in the x-plane can also be described by:

$$x = \frac{a \sin V}{\cosh U - \cos V}, \quad y = \frac{a \sinh U}{\cosh U - \cos V}. \tag{155}$$

V can be eliminated from this. With:

$$\frac{x}{y} = \frac{\sin V}{\sinh U} \tag{156}$$

and

$$\sin V = \sqrt{1 - (\cosh U - \frac{a}{y} \sinh U)^2} \tag{157}$$

we obtain:

$$x^2 + y^2 + a^2 - 2ay \coth U = 0 , \tag{158}$$

which can also be written:

$$x^2 + (y - a \coth U)^2 = \frac{a^2}{\sinh^2 U} . \tag{159}$$

For constant values of U, this is the equation of a circle with the center point on the y-axis.

U can also be eliminated. By the same computational scheme, there is obtained:

$$(x - a \cot V)^2 + y^2 = \frac{a^2}{\sin^2 V} \equiv r_0^2 . \tag{160}$$

The circles for constant V now have the center point on the x-axis. Because:

$$r_0^2 - x_0^2 = \frac{a^2}{\sin^2 V} - a^2 \cot^2 V = a^2 \tag{161}$$

they now intersect the imaginary axis at a distance of $\pm a$ from the origin; see Figure 39c. The circles from (159), on the other hand, do not intersect the real axis, as outlined in Figure 39b. The two equations (159) and (160) can also be expressed by polar coordinates r, ϑ. Substitution of

$$x = r \cos \varphi , \quad y = r \sin \varphi$$

gives:

$$\frac{r^2 + a^2}{2ar} \tanh U = \sin \psi , \tag{162}$$

and

$$\frac{r^2 - a^2}{2ar} \tan V = \cos \varphi .$$

(163)

If two different radii r and r' are now considered, and these are always chosen by the rule:

$$r\,r' = a^2 ,$$

(164)

then the following identity applies:

$$\frac{r^2 + a^2}{2ar} = \frac{r'^2 + a^2}{2ar'} .$$

(165)

If the two radii r, r' lie on the same ray defined by the angle ϑ then it can be deduced from (162) that they must also have the sum value of U after the transformation; see Figure 39d.

If the rule in (164) is observed, the following identity is also valid:

$$\frac{r^2 - a^2}{2ar} = - \frac{r'^2 - a^2}{2ar'}$$

(166)

It can then be concluded from (163), again if the two radii r, r' lie on the same ray fixed by the angle ϑ, that the tan values of V and V' must have opposite signs.

It is now desirable to complete the transition of the transformation formula in (150) to that in (142). If the w-components in the latter equation are designated as u, v, then from the comparison of:

$$U + iV = \ln \frac{z+ia}{z-ia}$$

$$u + iv = ih - \frac{h}{\pi} \ln \frac{z+ia}{z-ia}$$

(167)

there is immediately obtained:

$$U = -\frac{\pi}{h} u , \qquad V = \frac{\pi}{h} (h - v) .$$

(168)

If the value for V is substituted in (163), and (164) and (166) are still observed, then if δ signifies a specific value of v which is associated with the radius r, for which the value δ' is associated with the radius r', from:

$$\tan \frac{\pi}{h} (h - \delta) = - \tan \frac{\pi}{h} \delta = - \tan \frac{\pi}{h} (h - \delta') \qquad (169)$$

there results the equation:

$$\delta' = h - \delta , \qquad (170)$$

i.e., with the assumption of the condition in (164), there are produced values of v symmetrical with the center line.

The two equations (162) and (164) can now be combined by division, which results in:

$$\tan \varphi = \frac{r^2 + r\,r'}{r^2 - r\,r'} \cdot \frac{\tanh \dfrac{\pi}{h} u}{\tan \dfrac{\pi}{h} v} . \qquad (171)$$

The above equations can now be applied advantageously to transmission line cross sections whose contours in the x-y plane correspond to constant values of u or v in the w-plane.

Let us take as a first application the strip from Figure 15b perpendicular to a conductive plane, with the parameters of (92). If the strip is allowed to coincide with the real axis and the conductive plane with the imaginary axis of Figure 39a, then the application of the transformation produces a strip located transversely between two conductive planes (see Figure 48). Equation (149) can now be solved for r and the ratio r'/r can be formed:

Figure 40

We obtain:

$$\frac{r'}{r} = \frac{\tan \dfrac{\pi}{2} \dfrac{v_2}{h}}{\tan \dfrac{\pi}{2} \dfrac{v_1}{h}} . \tag{172}$$

By substitution in (92), it follows that:

$$k_A = \frac{\sqrt{\tan \dfrac{\pi}{2} \dfrac{v_2}{h} \Big/ \tan \dfrac{\pi}{2} \dfrac{v_1}{h} + 1}}{\sqrt{\tan \dfrac{\pi}{2} \dfrac{v_2}{h} \Big/ \tan \dfrac{\pi}{2} \dfrac{v_1}{h} - 1}} , \quad \lambda = \frac{1}{2} . \tag{173}$$
(in Table I, No. 36)

For the position symmetrical to the center, using $v_2 = h - v_1$ from (170) and with (89), there is produced the expression:

$$k_A = \cot\left(\frac{\pi}{4} - \frac{\pi v_1}{2h}\right) , \quad \lambda = \frac{1}{2} . \tag{174}$$
(in Table I, No. 35)

The transmission line with two transverse strips from Figure 40b produced by reflection at the real axis, finally, has the parameter k_A from (173) with $\lambda = 1$.

As the next easily transformed system, we suggest the circular segment above a conductive plane. In the case of the system of Figure 16b with the parameters of (85), a strip as in Figure 41a is formed by the transformation, parallel to the enveloping planes and lying in the center.

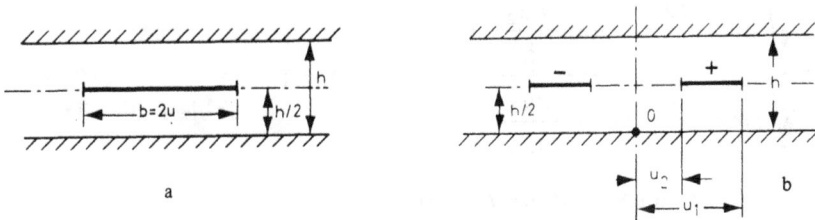

Figure 41

As a result of (146), for the half width u = b/2, there is immediately obtained:

$$u = \frac{h}{\pi} \ln \cot \left(\frac{\pi}{4} - \frac{\alpha}{2} \right). \tag{175}$$

Solved for $\alpha/2$ and introduced into (85), with (38), this leads to:

$$k_A = \coth \frac{\pi u}{2h}, \quad \lambda = \frac{1}{2}. \tag{176}$$

$$\text{(in Table I, No. 34)}$$

The system of two strips lying parallel between the planes as in Figure 41b is also of interest [19] and can be obtained by a transformation from two circular segments above a plane as in Figure 29a and the parameters of (120a). In this case, u_1 corresponds to the angle $\gamma + \alpha$, and u_2 corresponds to the angle $\gamma - \alpha$, in Figure 41b, and with (146), therefore:

$$u_1 = \frac{h}{\pi} \ln \cot \left(\frac{\pi}{4} - \frac{\gamma + \alpha}{2} \right),$$

$$u_2 = \frac{h}{\pi} \ln \cot \left(\frac{\pi}{4} - \frac{\gamma - \alpha}{2} \right). \tag{177}$$

Solution for $(\gamma + \alpha)$ and $(\gamma - \alpha)$, after application of the addition theorem:

$$\cot 2\alpha = \frac{\cot^2 \alpha - 1}{2 \cot \alpha} \tag{178}$$

then produces the parameters:

$$k_A = \frac{\sqrt{\sinh \dfrac{\pi u_1}{h} / \sinh \dfrac{\pi u_2}{h} + 1}}{\sqrt{\sinh \dfrac{\pi u_1}{h} / \sinh \dfrac{\pi u_2}{h} - 1}}, \quad \lambda = 1. \tag{179}$$

$$\text{(in Table I, No. 38)}$$

6. Three-Conductor Systems of Strips

In the discussion of three-conductor systems, two possible modes (even and odd) are to be differentiated [15, 21, 22] with the corresponding characteristic impedances z_{oe} and z_{oo}.

We would like to treat only one example of three coplanar strips in the following. Let the starting point be the coplanar three-strip transmission line of Figure 42a with symmetry at the zero point. Figure 42b shows the odd mode with the characteristic impedance z_{oo} to be determined, and Figure 42c the even mode with the impedance z_{oe} to be determined. It will be shown that the calculation of these impedances can be reduced to the case of two strip-lines and therewith to known results.

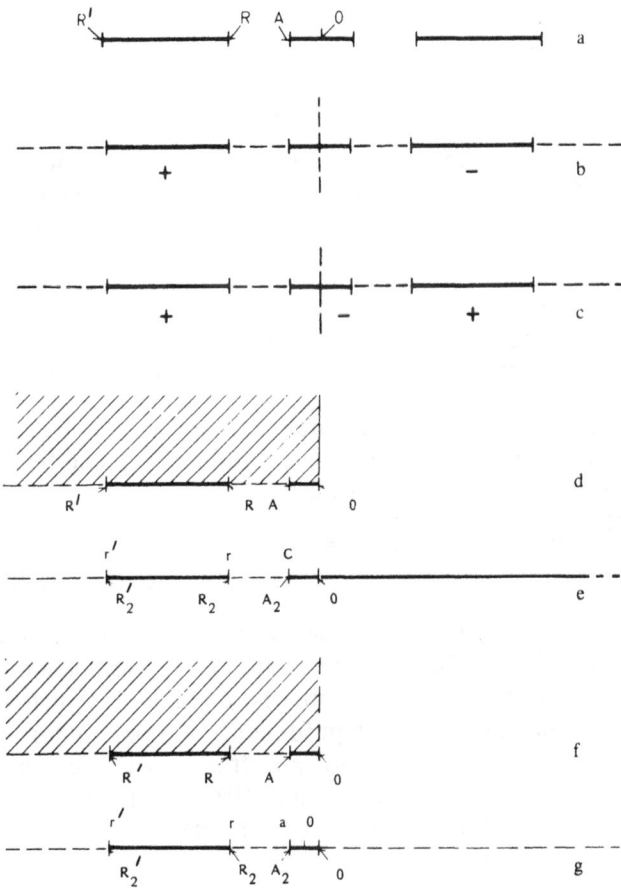

Figure 42

For the calculation of z_{oo}, we start from one quadrant, e.g., the upper left quadrant in Figure 42b, as drawn again in Figure 42d with its electrical and magnetic walls. It also has the characteristic impedance z_{oo}. By means of the transformation $w = Z^2$, this quadrant is depicted in the upper hemiplane of Figure 42e. Produced here are the equations:

$$A^2 = A_2 , \quad R^2 = R_2 , \quad R'^2 = R'_2 . \tag{180}$$

The characteristic impedance of this two-strip system in Figure 42e has already been determined over the entire plane by (132). The parameters r and r' or their ratio r/r' appear there. The following relationships apply:

$$r = R_2 - A_2 = R^2 - A^2 , \quad r' = R'_2 - A_2 = R'^2 - A^2 . \tag{181}$$

By introduction of these values into (132) and consideration of the factor of 2 for the hemiplane, we then obtain the parameters of the characteristic impedance z_{oo} for the three-strip system, as follows:

$$k_A = \frac{\sqrt[4]{(R'^2 - A^2)/(R^2 - A^2)} + 1}{\sqrt[4]{(R'^2 - A^2)/(R^2 - A^2)} - 1} , \quad \lambda = 1 . \tag{182}$$
(in Table I, No. 41)

The parameters for the complementary transmission line consisting of two identical strips in a slot in a conductive plane (voltage is between strips and plane) are found from this to be:

$$k_A = \sqrt[4]{(R'^2 - A^2)/(R^2 - A^2)} , \quad \lambda = 1 . \tag{183}$$
(in Table I, No. 43)

The calculation of z_{oe} takes place similarly. The starting point is the quadrant in Figure 42f with the characteristic impedance $4 Z_{oe}$, and it is portrayed in the upper hemiplane of Figure 42g. The equations in (180) again are produced. The parameters of this system of two strips of dissimilar widths over the entire plane (Figure 37b) are already known from (140). The parameters a, r', r appear there, or the sums $r'-a$; $r'+a$; $r-a$; $r+a$. The following relationships are valid:

$$a = A_2/2 , \quad r = R_2 - A_2/2 = R^2 - A^2/2, \quad r' = R'_2 - A_2/2 = R'^2 - A^2/2 , \tag{184}$$

$$r' - a = R'^2 - A^2 , \quad r' + a = R'^2 , \quad r - a = R^2 - A^2 , \quad r + a = R^2 .$$

By introduction of these values into (140) and consideration of
the factor 1/2, the parameters of the characteristic impedance
z_{oe} for the three-strip system are then found to be:

$$k_A = \frac{\left(\dfrac{R'^2 - A^2}{R'^2} \dfrac{R^2}{R^2 - A^2}\right)^{1/4} + 1}{\left(\dfrac{R'^2 - A^2}{R'^2} \dfrac{R^2}{R^2 - A^2}\right)^{1/4} - 1} \quad , \quad \lambda = \frac{1}{4} . \tag{185}$$

(in Table I, No. 42)

The parameters of the complementary transmission line consisting
of two identical strips in a slot in a conductive plane (voltage is
between the internal strips) are found from this to be:

$$k_A = \left(\frac{R'^2 - A^2}{R'^2} \frac{R^2}{R^2 - A^2}\right)^{1/4} \quad , \quad \lambda = 4 . \tag{186}$$

(in Table I, No. 44)

Chapter VII

Extension to Other Projection Groups (Special Strips, Several Thin Round Conductors, Parallel Strips)

1. Transformation Series

If no value is found at first for the characteristic impedance for a specific transmission line cross section, it is often suitable to undertake some transformations to establish to what group of systems the present case belongs. Even if in doing this, one does not run into a system already calculated, nevertheless, the advantage is generally gained of being able to seek out the most favorable system for the calculation. This will be illustrated briefly by an example.

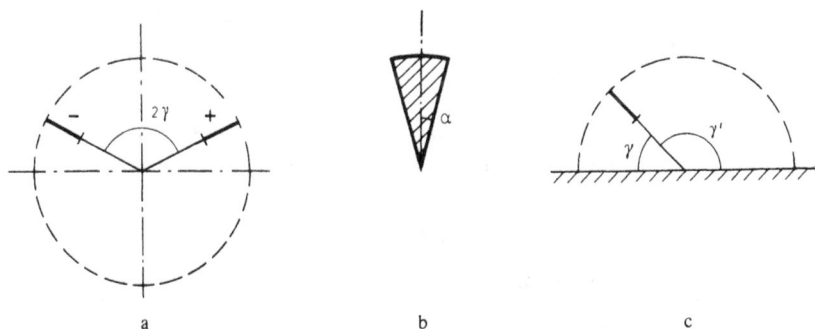

Figure 43

Assume, for example, a conical line such as that in Figure 43a to be determined, with two conical strips (Figure 43b) in different planes which intersect in a line perpendicular to the plane of the drawing passing through the origin. For reasons of symmetry, it is necessary to calculate only one-half as in Figure 43c, which consists of a conical strip which can be swiveled freely above a conductive plane. This system can now be projected onto the plane in various ways. If this is done in the position of Figure 44a, a segment of a circle is obtained in a circle as in Figure 44b.

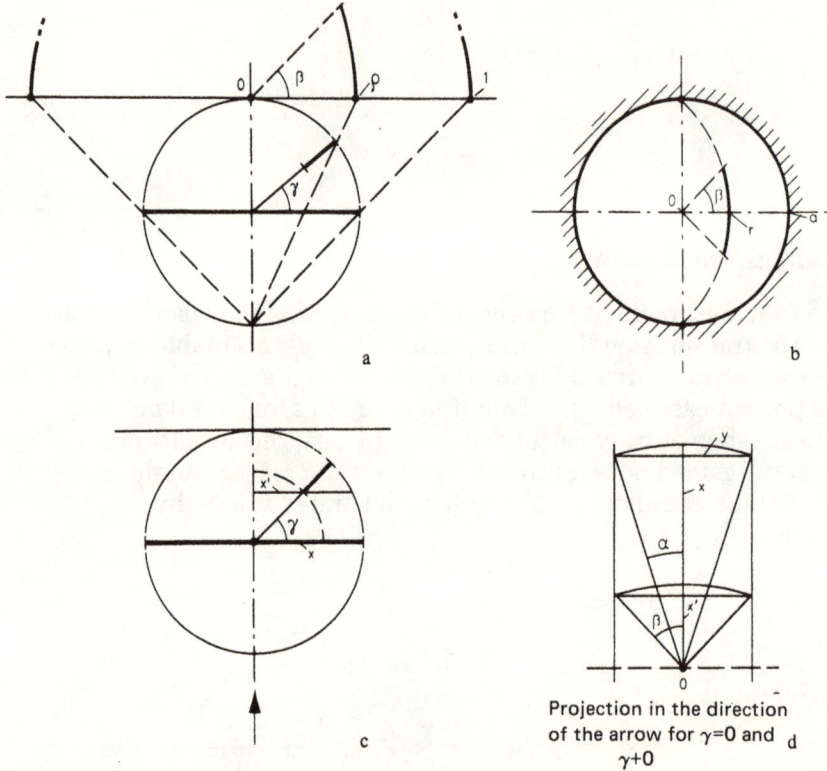

a

b

c

d

Projection in the direction
of the arrow for $\gamma=0$ and
$\gamma+0$

Figure 44

Right away, it can be deduced from this that:

$$\rho = \tan \frac{1}{2}\left(\frac{\pi}{2} - \gamma\right) = \frac{r}{a}.$$
(187)

However, the angle α of the conical strip changes into an angle of $\beta \geqslant \alpha$ in the plane. With the designations of Figure 44c, d, the following equations are deduced:

$$\frac{\tan \alpha}{\tan \beta} = \frac{x'}{x} \tag{188}$$

$$\frac{x'}{x} = \cos \gamma \tag{189}$$

Setting these equal, there is obtained:

$$\tan \alpha = \tan \beta \cdot \cos \gamma, \quad \text{bzw.} \quad \tan \alpha = \tan \beta \cdot \sin \gamma', \quad \frac{\pi}{2} - \gamma. \tag{190}$$

With (187) and (190), all necessary components of the transformation are determined.

If the sphere of Figure 44a is rotated around the axis perpendicular to the plane of the drawing by the angle π, a projection as in Figure 45a produces a circle outside of a circle, as in Figure 45b.

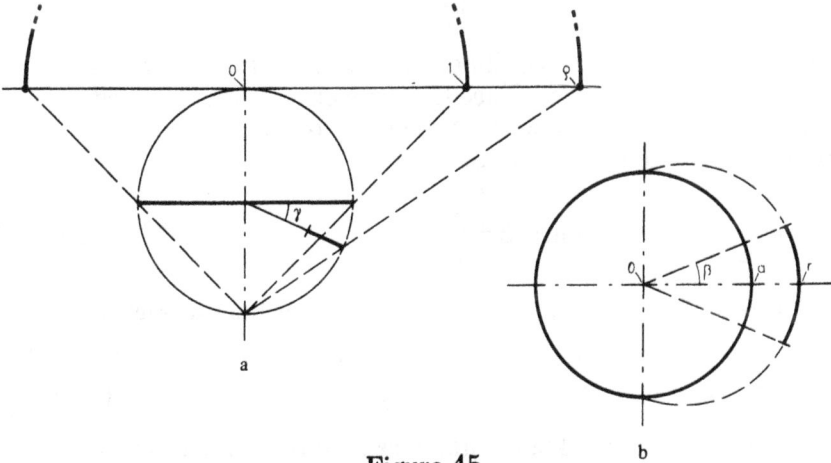

Figure 45

The following relationship is deduced:

$$\rho = \tan \frac{1}{2} (\pi + \gamma) = \frac{r}{a}. \tag{191}$$

In addition (190) still remains valid. If the sphere of Figure 44a is rotated around the axis perpendicular to the plane of the drawing by the angle $\pi/2$, a projection as in Figure 46a produces a segment of a circle opposite to a conductive plane, as in Figure 46b. It then follows that:

$$\rho = \tan \gamma'/2 = \frac{r}{a}. \tag{192}$$

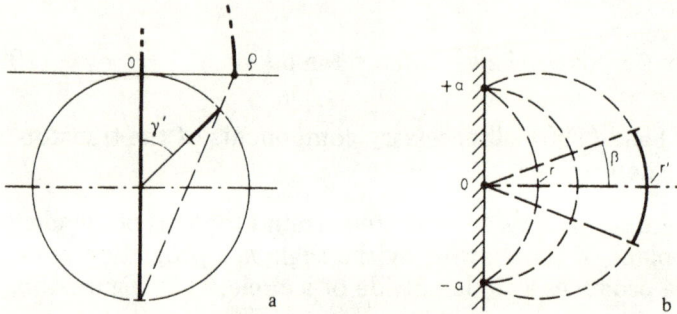

Figure 46

In combination with (190), all necessary transformation equations are therewith determined. Since the same characteristic impedance of the conical stripline must result for the angle $(\pi - \gamma)$, this also applies to the projection, because:

$$\rho' = \tan \frac{1}{2} \; (\pi - \gamma) = \cot \gamma/2 = \frac{r'}{a} \tag{193}$$

the following equation applies to the equivalent lines in Figure 46b:

$$r \cdot r' = a^2 . \tag{194}$$

If the sphere of Figure 44a is rotated around the horizontal axis in the plane of the drawing by the angle $\pi/2$, a projection as in Figure 47a produces a strip lying in a plane which passes through the zero point, opposed to a conductive plane, as in Figure 47b.

From this it follows that:

$$\rho = \tan \frac{1}{2}\left(\frac{\pi}{2} - \alpha\right) = \frac{r}{a} = \sqrt{\frac{r}{r'}} . \tag{195}$$

The angle γ in this projection remains unchanged.

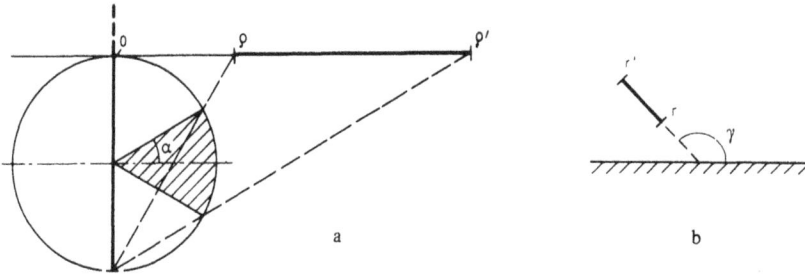

Figure 47

The system of Figure 47b can now be subjected again to the transformation $w = z^n$. From the original parameters designated by the index 1, the following are then derived:

$$r_2 = r_1^n, \quad r_2' = r_1'^n, \quad \gamma_2 = n\gamma_1. \qquad (196)$$

Depending upon the choice of n, a strip opposed to an acute angle as in 48a is then produced for $1 < n < 2$, a strip opposed to the edge of a conductive plane as in 48b for $n = 2$, and a strip in a reentrant angle, Figure 48c, for $0 < n < 1$. From these, strips with the same distance from the origin and on straight lines passing through it can be produced by reflection, as in Figure 49a.

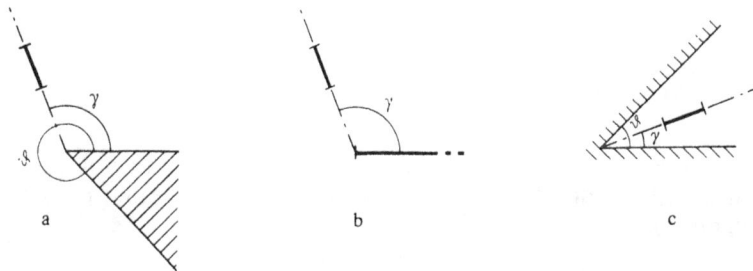

Figure 48

A transformation of the systems in Figure 48 onto a sphere leads to a conical strip opposed to a wedge as in Figure 49b, and to a conical strip opposed to the edge of a conductive plane as in Figure 49c, while the transformation of the system in Figure 49a leads to the corresponding multi-strip conical transmission lines. Only the systems in Figure 50a-h will be pointed out as examples of the other line cross sections to be obtained from Figure 49 by transformation onto a plane.

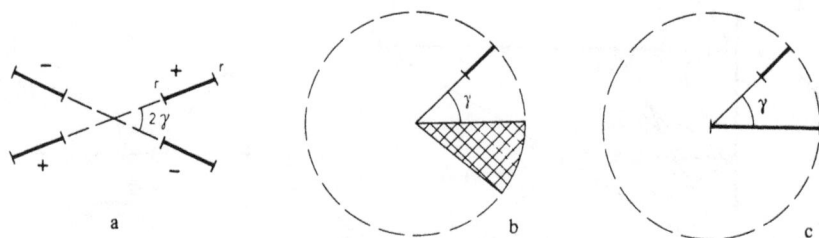

Figure 49

By representing a half-space on a strip, parallel strip conductors can again be obtained. From the segment of a circle lying on a line of constant value of v in the x-y plane, at an arbitrary distance from the conductive plane of Figure 46b, the strip of Figure 51a can be obtained at any arbitrary distance from, and parallel to, two enclosing planes.

The parameters v and u are determined from β and r through (149), (162), and (168) to be:

$$v = h \left(1 - \frac{2}{\pi} \text{ arc tan } \frac{a}{r} \right) \qquad\qquad (197)$$

and

$$u = \frac{h}{\pi} \text{ ar tanh } \left(\frac{2ar}{r^2 + a^2} \sin \beta \right) . \qquad\qquad (198)$$

Finally, reflection at one conductive plane of Figure 51a obtains the transmission line of two parallel strips between parallel planes as in Figure 51b.

After various transformations, therefore, a whole series of new systems has been produced from the conical stripline with different axial directions as in Figure 43. So far, besides the special case $\gamma = \pi/2$, no connection has been discovered to any of the previously treated and calculated systems. In this situation, one can proceed to seek out some systems suitable for the calculation. Presumably, the system in Figure 51a would be best suited here, of which one-half can be portrayed in a hemiplane according to Schwarz-Christoffel, and can there be brought into the form of one of the transmission lines in Figure 20b, 15c, 34b, or 37b. (see also Hachemeister, ref. [37].)

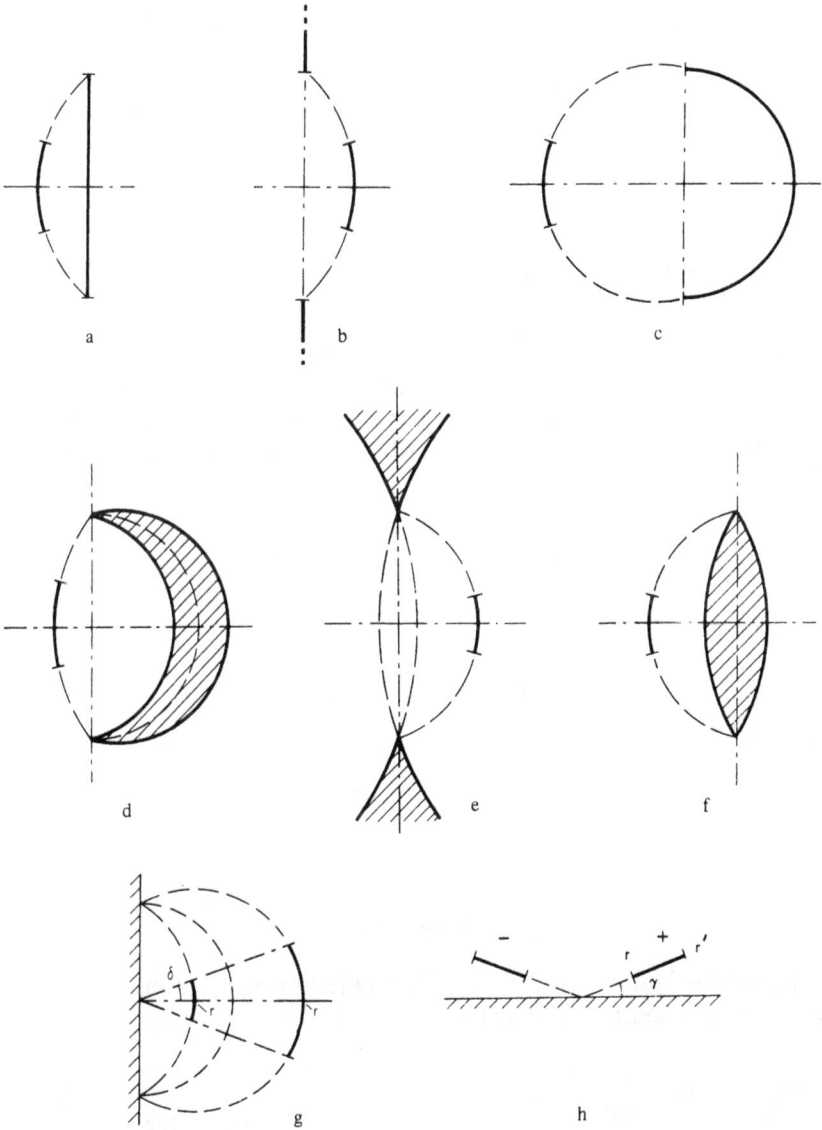

Figure 50

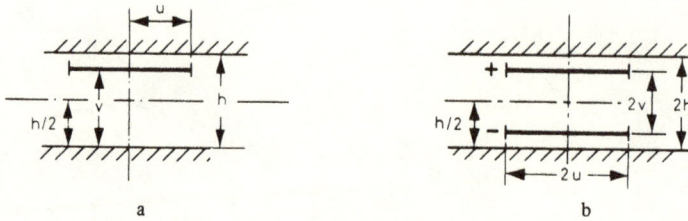

Figure 51

2. *Thin Round Conductors and Strips, Push-Pull*

The multi-conductor systems treated in the earlier literature consist almost exclusively of circular conductors whose diameters are small in comparison with the distance to other circular conductors, planes, or strips. To supplement this, it should also be shown here that the known results can also be obtained very easily with the use of the calculation process discussed.

The characteristic impedance of a circular cylindrical conductor opposite a conductive plane in Figure 10b is given exactly by (60).

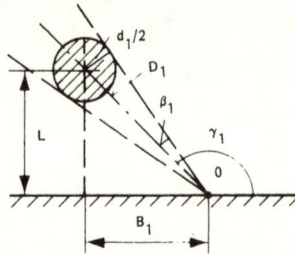

Figure 52

Using the designations of Figure 52, the length parameters L, d can also be expressed by angular parameters γ_1, β_1 :

$$Z = \frac{\eta}{2\pi} \text{ ar cosh } \frac{\sin \gamma_1}{\sin \beta_1} . \qquad (199)$$

(in Table I, No. 6)

If this system is subjected to the transformation $w = z^n$ pursuant to (113), then the conductor cross section transformed from the circular cross section, which is no longer a circle because of the non-linear transformation, turns out to lie in the angular region

$n(\gamma_1 + \beta_1)$ and $n(\gamma_1 - \beta_1)$. If the diameter of the original circle
was small in comparison with the distance to the origin, however,
a circle will also be formed to a good approximation after the
transformation. Proceeding from this assumption, for the assumption, for the transformation with $1 < n < 2$, a circle opposed to
an acute angle is obtained as in Figure 53a, while on the other hand
with $n = 2$, a circle opposed to the edge of a conductive plane is
obtained as in Figure 53b, and with $0 < n < 1$, a circle is obtained
in a reentrant angle, as in Figure 53c. From (199), we then have:

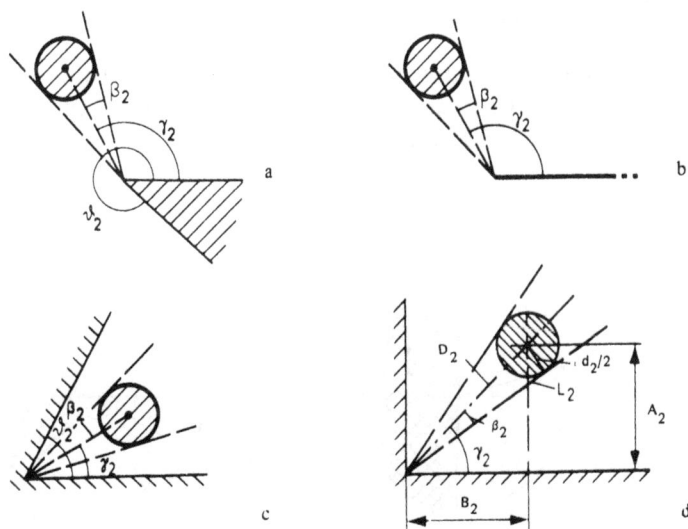

Figure 53

$$Z_h = \frac{\eta}{2\pi} \text{ ar cosh } \frac{\sin \dfrac{\gamma_2}{n}}{\sin \dfrac{\beta_2}{n}}, \qquad (200)$$

and if it is also considered that for the angle ϑ_2, π must again be
produced by a retransformation (115), this then leads to the
equation:

$$Z_h = \frac{\eta}{2\pi} \text{ ar cosh } \frac{\sin \pi \dfrac{\gamma_2}{\vartheta_2}}{\sin \pi \dfrac{\beta_2}{\vartheta_2}}, \quad \vartheta_2 - \beta_2 > \gamma_2 > \beta_2. \qquad (201)$$

(in Table I, No. 47)

In particular, for the round conductor opposite the edge of a conductive plane with $\vartheta_2 = 2\pi$, we obtain:

$$Z_h = \frac{\eta}{2\pi} \text{ ar cosh } \frac{\sin \gamma_2/2}{\sin \beta_2/2} , \qquad\qquad (202)$$

(in Table I, No. 48)

and for the circular conductor in a rectangular corner with $\vartheta_2 = \pi/2$ in Figure 53d:

$$Z_h = \frac{\eta}{2\pi} \text{ ar cosh } \frac{\sin 2\gamma_2}{\sin 2\beta_2} . \qquad\qquad (203)$$

(in Table I, No. 49)

Because of the relatively large arguments of the arcosh function, the ln function can always be used instead of it here and in the following (see Section VIII.2).

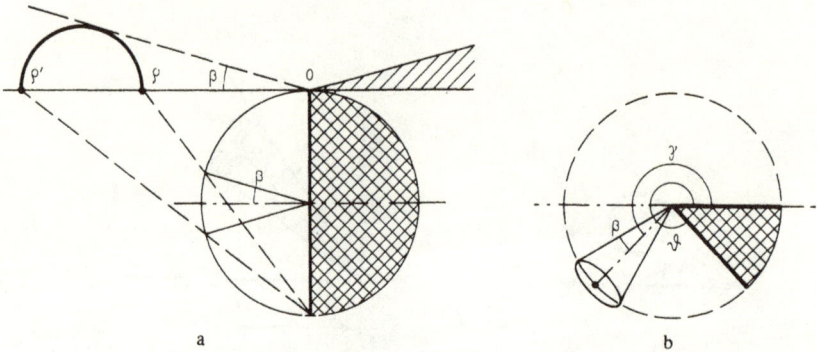

Figure 54

By a projection of these flat systems from Figure 53 onto a sphere in such a way that the origin in the plane coincides with the North Pole of the sphere as in Figure 54a, a circular cone opposite a wedge is obtained in Figure 54b, or opposite the edge of a plane. The angles do not change in this projection, so that Equations (201), (202), (203) apply here also. From these circular cone-wedge systems can now be obtained new parallel lines by transformation onto a plane. If the circular cone opposed to the edge of a conductive plane in the position of 55a is projected onto a plane, a circular conductor opposed to an open circle is obtained, as in Figure 55b.

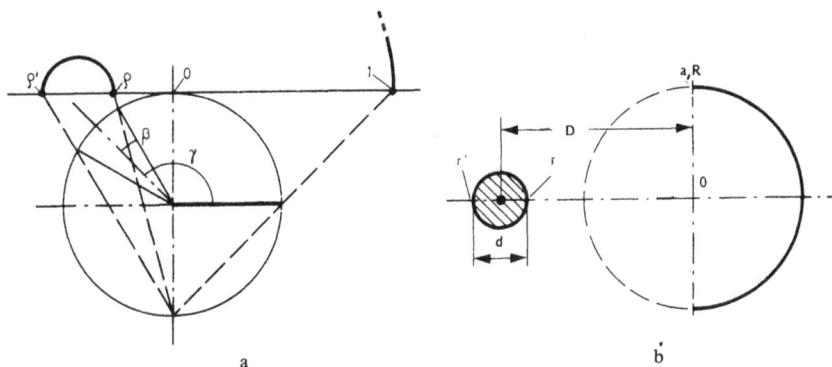

Figure 55

From this, it follows that

$$\rho = \tan \frac{1}{2}\left(\gamma_2 - \beta_2 - \frac{\pi}{2}\right) = \frac{r}{a} \ , \quad \rho' = \tan \frac{1}{2}\left(\gamma_2 + \beta_2 - \frac{\pi}{2}\right) = \frac{r'}{a}. \tag{204}$$

Solved for $\gamma_2/2$ and $\beta_2/2$, and introduced into (202), with (18) this provides:

$$Z_h = \frac{\eta}{2\pi} \text{ ar cosh } \frac{\sin\left(\dfrac{\pi}{4} + \dfrac{1}{2} \text{ arc tan } \dfrac{a(r+r')}{a^2 - rr'}\right)}{\sin\left(\dfrac{1}{2} \text{ arc tan } \dfrac{a(r'-r)}{a^2 + rr'}\right)}. \tag{205}$$

(in Table I, No. 50)

With the parameters:

$$d = r' - r, \quad R = a, \quad 2D = r + r' \tag{206}$$

and with the use of (47), this can also be written as follows:

$$Z_h = \frac{\eta}{2\pi} \text{ ar cosh } \frac{\sin\left(\dfrac{\pi}{4} + \dfrac{1}{2} \text{ arc tan } \dfrac{8RD}{4R^2 - 4D^2 + d^2}\right)}{\sin\left(\dfrac{1}{2} \text{ arc tan } \dfrac{4Rd}{4R^2 + 4D^2 - d^2}\right)} \tag{207}$$

(in Table I, No. 50)

If the sphere in Figure 55a is rotated around the axis perpendicular to the plane of the drawing in a positive direction by the angle $\pi/2$, as in Figure 56a a circular wire opposed to a parallel strip is produced, Figure 56b.

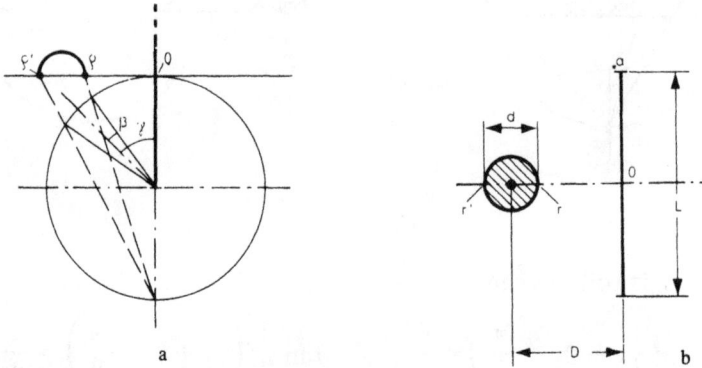

Figure 56

From this it can be shown that:

$$\rho = \tan \frac{1}{2}(\gamma_2 - \beta_2) = \frac{r}{a}, \quad \rho' = \tan \frac{1}{2}(\gamma_2 + \beta_2) = \frac{r'}{a}. \tag{208}$$

Solution for $\gamma_2/2$ and $\beta_2/2$ and introduction into (202) again with (18) leads to:

$$Z_h = \frac{\eta}{2\pi} \text{ ar cosh } \frac{\sin\left(\dfrac{1}{2} \text{ arc tan } \dfrac{a(r+r')}{a^2 - rr'}\right)}{\sin\left(\dfrac{1}{2} \text{ arc tan } \dfrac{a(r'-r)}{a^2 + rr'}\right)} \tag{209}$$
(in Table I, No. 51)

If the parameters of Figure 56b

$$L = 2a, \quad 2D = r + r', \quad d = r' - r, \tag{210}$$

are used, then we can also write:

$$Z_h = \frac{\eta}{2\pi} \text{ ar cosh } \frac{\sin\left(\dfrac{1}{2} \text{ arc tan } \dfrac{4LD}{L^2 - 4D^2 + d^2}\right)}{\sin\left(\dfrac{1}{2} \text{ arc tan } \dfrac{2LD}{L^2 + 4D^2 - d^2}\right)} \tag{211}$$
(in Table I, No. 51)

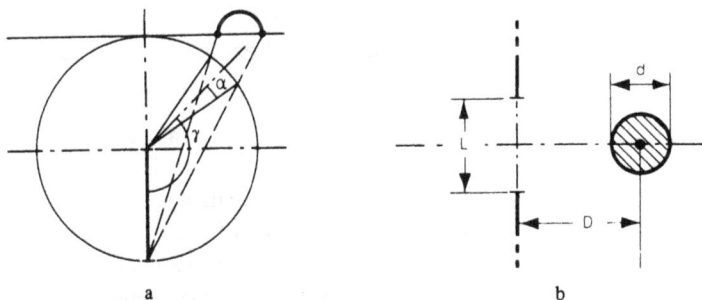

Figure 57

Finally, the sphere in Figure 55a can be rotated around the axis perpendicular to the plane of the drawing in the negative direction by the angle $\pi/2$, and projection onto a plane as in Figure 57a then produces a circular wire in the gap in a conductive plate, Figure 57b. From this it can be shown that:

$$\rho = \tan\frac{1}{2}\left(\gamma_2 + \beta_2 - \frac{\pi}{2}\right) = \frac{r}{a},$$

$$\rho' = \tan\frac{1}{2}\left(\gamma_2 - \beta_2 - \frac{\pi}{2}\right) = \frac{r'}{a}. \tag{212}$$

Solution for $\gamma_2/2$ and $\beta_2/2$, and introduction into (202), with (18) and the parameters in (210) leads to:

$$Z_h = \frac{\eta}{2\pi}\,\text{ar cosh}\,\frac{\sin\left(\frac{\pi}{2} - \frac{1}{2}\,\text{arc tan}\,\frac{4LD}{L^2-4D^2+d^2}\right)}{\sin\left(\frac{1}{2}\,\text{arc tan}\,\frac{2LD}{L^2+4D^2-d^2}\right)}. \tag{213}$$

(in Table I, No. 52)

In particular, with the equation obtained from (18):

$$\frac{1}{2}\,\text{arc tan}\,\frac{2u}{1-u^2} = \text{arc tan}\,u \tag{214}$$

it follows from (213) that the characteristic impedance for the circular conductor located precisely in the center of the system, and therefore for $D = 0$, is:

$$Z_h = \frac{\eta}{2\pi}\,\text{ar cosh}\,1/\sin\,\text{arc tan}\,(d/L) \approx \frac{\eta}{2\pi}\,\ln\frac{2L}{d}. \tag{215}$$

The result is suitable for comparison with the earlier results from Figure 14 and (90). It can be seen that because $L/d = r'/r$ for small values of d/L, the two equations convert into one another.

The general cases of the projection of the circular cone-wedge system of Figure 54b in various positions, will not be further discussed here for the sake of brevity. Furthermore, they obviously lead initially to structures of no technical interest. On the other hand, it is worthwhile to follow up to some extent the circular conductor in a rectangular corner as in Figure 53d and (203). First, length parameters will be introduced again instead of the angular parameters. In view of the following relationships:

$$\sin \beta = d_2/2D_2 , \qquad \cos \beta = L_2/D_2 \approx 1,$$

$$\sin \gamma = A_2/D_2 , \qquad \cos \gamma = B_2/D_2 , \quad D_2 = \sqrt{A_2{}^2 + B_2{}^2} , \qquad (216)$$

with the parameters A_2, B_2, L_2, and d_2 indicated in Figure 53d and after application of the addition theorem:

$$\sin 2\alpha = 2 \sin \alpha \cos \alpha \qquad\qquad (217)$$

the characteristic impedance is obtained in the form:

$$Z_h = \frac{\eta}{2\pi} \text{ ar cosh } \frac{2A_2 B_2}{d_2 \sqrt{A_2{}^2 + B_2{}^2}} \qquad\qquad \begin{matrix}(218)\\ \text{(in Table I, No. 49)}\end{matrix}$$

If the circular conductor is now reflected on the vertical plane, we obtain the symmetrical transmission line above a conductive plane as in Figure 58a. (The conductor diameters here and in the following are shown in exaggerated size to facilitate dimensioning, with the assumption, as always, that the diameters of the conductors should be small in comparison with their separation.)

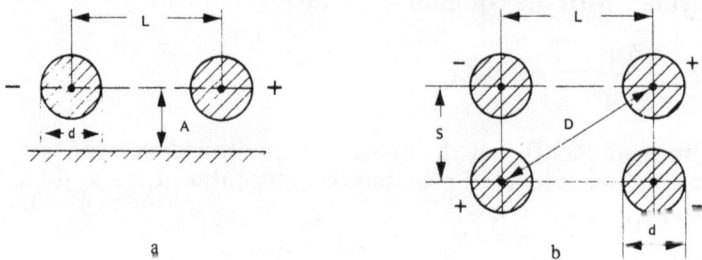

Figure 58

The characteristic impedance is double that of (218) with $L = 2B_2$, or:

$$Z_h = \frac{\eta}{\pi} \text{ ar cosh } \frac{A_2 L}{d_2 \sqrt{A_2{}^2 + L^2/4}} . \qquad\qquad (219)$$

(in Table I, No. 53)

If the two wire system is further reflected on the horizontal plane, the four-wire line of Figure 58b is obtained. The characteristic impedance amounts to one-half of (219), with $S = 2A_2$, or:

$$Z_h = \frac{\eta}{2\pi} \text{ ar cosh } \frac{SL}{d\sqrt{S^2 + L^2}} = \frac{\eta}{2\pi} \text{ ar cosh } \frac{SL}{dD} . \qquad (220)$$

(in Table I, No. 54)

For approximation calculation of the case in which thin circular conductors are distributed uniformly on the circumference of a circle, the transformation $w = z^n$ can also be used. This point is reached by proceeding from the circular conductor above a plane as in Figure 52 and (199) with $\gamma_1 = \pi/2$, after transformation into a reentrant angle as in Figure 59a, which produces the equation:

$$Z_h = \frac{\eta}{2\pi} \text{ ar cosh } 1/\sin \pi \frac{\beta_2}{\vartheta_2} . \qquad\qquad (221)$$

If ϑ_2 is now selected with the following limitation, where n signifies a whole number:

$$N \cdot \vartheta_2 = \pi , \quad N \geqslant 2 , \qquad\qquad (222)$$

then N acute angles exactly fill a hemiplane. Because of the reflection principle, with alternating polarity of the conductors in Figure 59b, the removal of the separating walls has no effect, so that a basketlike conductor is formed as in Figure 59c. The characteristic impedance of this is then:

$$Z_h = \frac{\eta}{N\pi} \text{ ar cosh } 1/\sin N\beta_2 . \qquad\qquad (223)$$

Since Figure 59a shows that:

$$\sin \beta_2 = d_2/2D_2 , \qquad\qquad (224)$$

with $D - 2D_2$, we can also write

$$Z_h = \frac{\eta}{N\pi} \text{ ar cosh } 1/\sin(N \text{ arc sin } (d/D)) \approx \frac{\eta}{N\pi} \ln \frac{2D}{Nd}. \qquad (225)$$

(in Table I, No. 55)

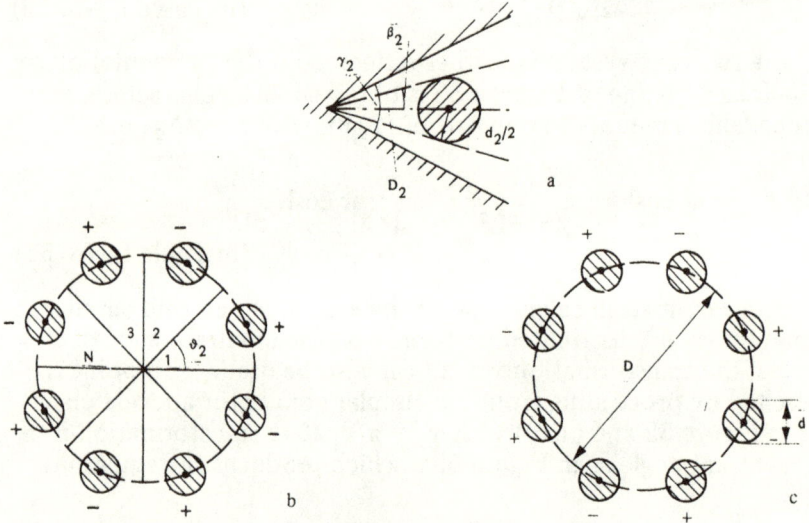

Figure 59

If the planar multi-conductor systems are portrayed on a sphere, a large number of planar systems can be generated from the conical systems formed there, again by projection onto a plane. At this point, the technically important shielded 2-wire transmission line might be treated only briefly as an example.

The symmetrical line above a conductive plane of Figure 58a and (219), in the notation with angular parameters as in (203), can be portrayed on a sphere, and in this way are obtained two identical circular cones above a conductive plane, whose axes in the simplest case can lie in a plane perpendicular to the base plane, where they are symmetrical to the central perpendicular. (However, circular cones can also be obtained which are not perpendicular to the conductive plane.) If a projection is then again undertaken in accordance with Figure 60a onto the plane, two circular conductors are obtained in a round shield, Figure 60b. From this, It follows that:

$$\rho = \tan \frac{1}{2}\left(\frac{\pi}{2} - \gamma - \beta\right) = \frac{r}{a},$$
$$\left.\begin{array}{r} \\ \\ \end{array}\right\}$$
$$\rho' = \tan \frac{1}{2}\left(\frac{\pi}{2} - \gamma + \beta\right) = \frac{r'}{a}.$$

(226)

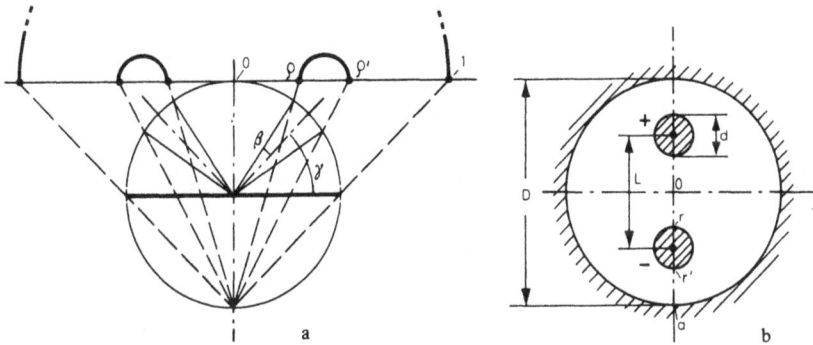

Figure 60

Solving for γ and β and introduction into (203), observing (18) and the factor of 2, then gives the characteristic impedance:

$$Z_h = \frac{\eta}{\pi} \text{ ar cosh} \frac{\sin 2 \text{ arc tan} \dfrac{a(r'+r)}{a^2 - rr'}}{\sin 2 \text{ arc tan} \dfrac{a(r'-r)}{a^2 + rr'}}.$$

(227)
(in Table I, No. 56)

In Figure 60a, if cones which are not perpendicular to a conductive plane are used, internal conductors are produced in Figure 60b which are displaced from the perpendicular line of symmetry, etc. In the same way, shielded cables with more than two internal conductors can be calculated. The accuracy here depends only on how well the characteristic impedance of the conductor above a conductive plane was determined.

Some of the transmission lines treated in this section, particularly those in Figure 60b, have been calculated in the literature in another way and are known with great precision [30]. In case of higher

demands, therefore, it is much better to begin with the computa-
tions at this point, and to retrace the transformation series. However, this will not be carried out any further at this time, since the
main concern here has been to set forth the principle of the devel-
opment of a transformation series and to carry it to the point at
which suitable systems for calculation are eventually found.

3. Thin Round Conductors, Synchronous

The transmission line formed of two circular conductors above a
conductive plane with synchronous excitation as in Figure 61a can
apparently not be obtained by a transformation from the previous
systems. However, if the diameters are small in comparison with
the separations, the characteristic impedance can be approximated
easily.

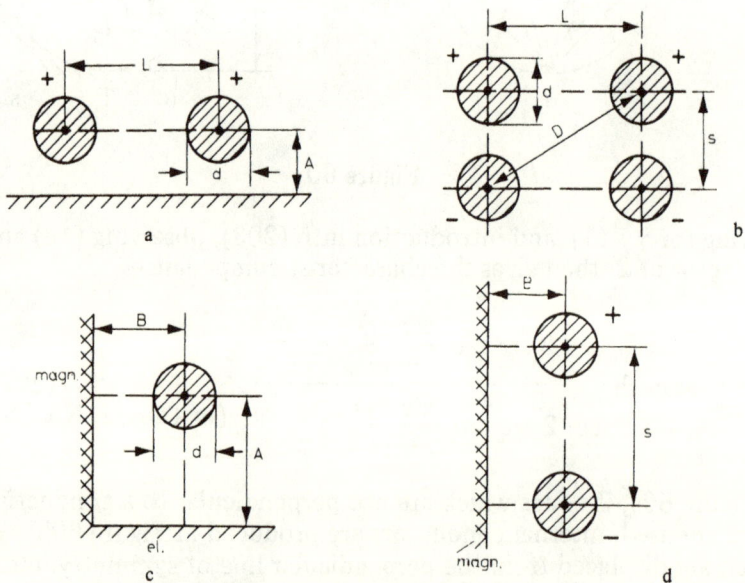

Figure 61

For this purpose, one goes back to line charges, as in the system
of Figure 62a (the upper two conductors in Figure 61a should be
imagined reflected on a horizontal plane). First, however, the
supposition should be confirmed that the equipotential lines at
a small distance from the line charges are circular to a good
approximation.

If the charge density is designated as q, its potential φ at a distance r is:

$$\varphi = - \frac{q}{2\pi\epsilon} \ln r + \text{const.} \tag{228}$$

For two homonymous line sources as in Figure 62b, the potential at a point in space P is found to be:

$$\varphi = - \frac{q}{2\pi\epsilon} \ln r_1 r_2 , \tag{229}$$

wherein the lengths r_1 and r_2 are:

$$r_1 = \sqrt{x^2 + y^2} , \quad r_2 = \sqrt{(x+L)^2 + y^2} . \tag{230}$$

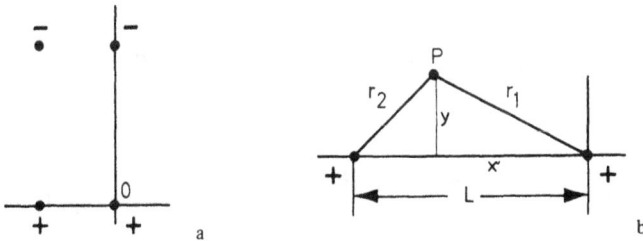

Figure 62

For lines of constant potential φ, which are therefore possible conductor surfaces, if K is an arbitrary constant, then because of (229), the following must apply:

$$(r_1 \cdot r_2)^2 = (x^2 + y^2)((x+L)^2 + y^2) = k^2 . \tag{231}$$

If the neighborhood of the right line source is considered in a region x, y $\ll$ L, then to a good approximation:

$$x^2 + y^2 = \frac{k^2}{L^2} = \frac{d^2}{4} , \tag{232}$$

i.e., in the neighborhood of this line source the equipotential surfaces are actually cylindrical to a first approximation, with the line source as the axis. For reasons of symmetry, this must then also be true for the left line source, and also for other line sources with large separations from one another, as for example in Figure 62a.

Considering this result, the characteristic impedance of four circular wires in the synchronous excitation of Figure 61b will now be calculated. The voltage U between the upper and lower wires is the same as the potential difference of the right lower and the right upper conductors, for example, or:

$$U = - \frac{q}{2\pi\epsilon} \ln \sqrt{\frac{\frac{d^2}{4}\left(L - \frac{d}{2}\right)^2}{\left(S - \frac{d}{2}\right)^2\left(\left(S - \frac{d}{2}\right)^2 + L^2\right)}} \approx$$

$$- \frac{q}{2\pi\epsilon} \ln \frac{dL}{2S\sqrt{S^2 + L^2}} \quad . \tag{233}$$

From this, with (2) and the simplification $D = \sqrt{S^2 + L^2}$, the characteristic impedance turns out to be:

$$Z_h = - \sqrt{\frac{\mu}{\epsilon}} \frac{1}{2\pi} \ln \frac{dL}{2SD} = \frac{\eta}{2\pi} \ln \frac{2SD}{dL} \quad . \tag{234}$$
$$\text{(in Table I, No. 58)}$$

For comparison with the push-pull excitation in (220), equal distances $S = L$ are best chosen, for which the result is:

$$Z_h = \frac{\eta}{2\pi} \ln \frac{\sqrt{2}\,L}{d} \quad , \qquad \text{Push-pull } (Z_{oo}) \tag{235a}$$

$$Z_h = \frac{\eta}{2\pi} \ln \frac{2\sqrt{2}\,L}{d} \quad , \qquad \text{Synchronous } (Z_{oe}) \tag{235b}$$

For the same ratio L/d as expected, the characteristic impedance in synchronous excitation is somewhat greater than in push-pull excitation [15].

The characteristic impedance of the two circular conductors above a conductive plane as in Figure 61a is one-half the value in (234), or with $S/2 = A$:

$$Z_h = \frac{\eta}{4\pi} \ln \frac{4A\sqrt{L^2 + 4A^2}}{dL} \quad . \tag{236}$$
$$\text{(in Table I, No. 59)}$$

The characteristic impedance of the circular conductor in the rectangular corner consisting of one electrical and one magnetic wall as in Figure 61c is double that of (236), or with $L/2 = B$:

$$Z_h = \frac{\eta}{2\pi} \ln \frac{4A\sqrt{A^2+B^2}}{dB}, \qquad\qquad (237)$$

(in Table I, No. 60)

while the characteristic impedance of the symmetrical transmission line opposed to the magnetic plane as in Figure 61d is again double that of (237):

$$Z_h = \frac{\eta}{\pi} \ln \frac{S\sqrt{S^2+4B^2}}{dB}. \qquad\qquad (238)$$

(in Table I, No. 61)

The system in Figure 61d can now be well chosen as the starting point for further transformations, for example, in order to estimate the effect of magnetic envelopes or additional magnetic bodies on the characteristic impedance. However, this will not be carried out here.

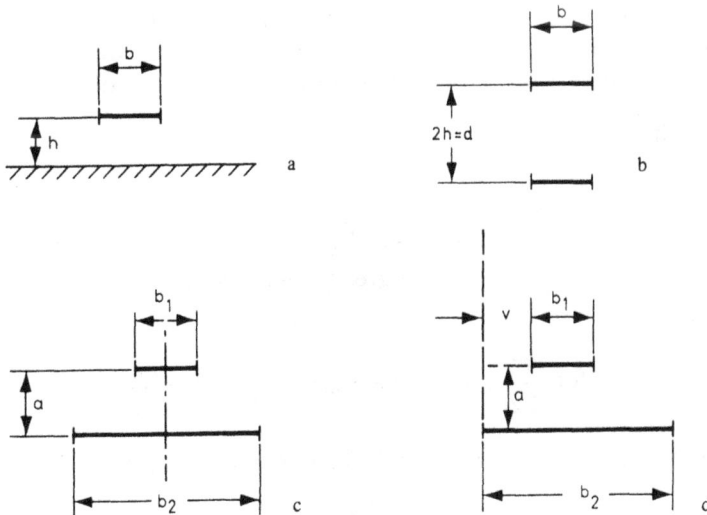

Figure 63

4. The Parallel Stripline (Microstrip) and Related Lines

A Transmission line comparatively inaccessible to the "exact" calculation is the unshielded parallel stripline in the various embodiments illustrated in Figures 63a, b, c, d. Magnus and

Oberhettinger, first of all, have carried out an exact calculation for the lines in Figure 63a, b [12]. However, the analytical solution is so unwieldy that it has been attempted again and again to derive useful approximations for very narrow and very broad strips, i.e., for b ≪ h and b ≫ h. Let us consider some approximations in some detail:

The case of narrow bands follows from the system in Figure 16b. If the length of the circular segment is designated as s and the radius as h, then:

$$s = 2h\alpha .$$ (239)

For α, this system then approximates that in Figure 63a with b, so that by introduction of the value α from (239) into (85) for $s \cong b$, we obtain:

$$Z_h = \frac{\eta}{2\pi} \ln \left(2 \cot \frac{b}{4h} \right) \approx \frac{\eta}{2\pi} \ln \frac{8h}{b} .$$ (240)
(in Table I, No. 62)

For the case of broad bands, to a first approximation, the known capacity of plate capacitors could be used, ignoring the boundary effects, whereby the characteristic impedance turns out to be:

$$Z_1 \approx \eta \frac{h}{b} .$$ (241)

However, this formula can be used only for the most rough estimations. There are some with greater accuracy. One of these reads [162]:

$$\frac{h}{b} = \pi/2 \, (\ln\cosh (\frac{\pi\eta}{2Z}) - \ln [\ln \cosh (\frac{\pi\eta}{2Z})] - 1)$$ (242)

For wide bands, i.e., for $\pi\eta/2Z \gg 1$, an excellent approximation is:

$$\frac{h}{b} = \pi/2 \, (\frac{\pi\eta}{2Z} - \ln 2 - \ln (\frac{\pi\eta}{2Z} - \ln 2) - 1)$$

If the approximations for narrow and wide bands are now combined, and the mutual boundary is chosen at h/b = 1.1, the entire possible range of values can be covered with a maximum error of only 2.7%, which occurs precisely at this boundary (in [162], the error had been estimated to be too good by a factor 1/4). It then drops rapidly to zero even at a relatively short distance. In summary, it follows that:

$$\frac{h}{b} = \pi/2 \left(\frac{\pi\eta}{2Z_u} - \ln 2 - \ln \left(\frac{\pi\eta}{2Z_u} - \ln 2 \right) - 1 \right), \tag{243a}$$

$$0 \leqslant \frac{h}{b} \leqslant \infty, \quad 0 \leqslant Z_u/\eta \leqslant 0.35$$

$$\frac{h}{b} = \frac{1}{8} e^{Z_u 2\pi/\eta}, \quad 1.1 \leqslant \frac{h}{b} \leqslant \infty, \quad 0.35 \leqslant Z_u/\eta \leqslant \infty \tag{243b}$$
(in Table I, No. 62)

The index u on Z indicates the asymmetrical transmission line of Figure 63a.

By reflection of the system in Figure 63a at the conducting plane, the symmetrical transmission line of Figure 63b is obtained, consisting of identical parallel strips with the characteristic impedance formula ($Z_s = 2Z_u$, $d = 2h$)

$$\frac{d}{b} = \pi/ \left(\frac{\pi\eta}{Z_s} - \ln 2 - \ln \left(\frac{\pi\eta}{Z_s} - \ln 2 \right) - 1 \right), \quad 2.2 \leqslant \frac{d}{b} \leqslant \infty, \tag{244a}$$

$$0 \leqslant Z_s/\eta \leqslant 0.7,$$

$$\frac{d}{b} = \frac{1}{4} e^{Z_s \pi/\eta}, \quad 0 \leqslant \frac{d}{b} \leqslant 2.2, \quad 0.7 \leqslant Z_s/\eta \leqslant \infty. \tag{244b}$$
(in Table I, No. 63)

In order to discover some closely related transmission lines, we can first determine again the equivalent conical transmission line. If the planar cross section of Figure 63a is projected onto a sphere as in Figure 64a, there is formed a segment of a hollow cone opposed to a conductive plane, as outlined in perspective in Figure 64b. To obtain the projection equations, some supplementary parameters m, n, z, β, d are drawn into Figures 64c, d, for which it can be deduced that:

$$n^2 = 1 + (h/a)^2, \quad \tan\beta = \frac{b/2a}{n}, \quad \frac{h}{a} = \tan\frac{\alpha_1}{2}. \tag{245}$$

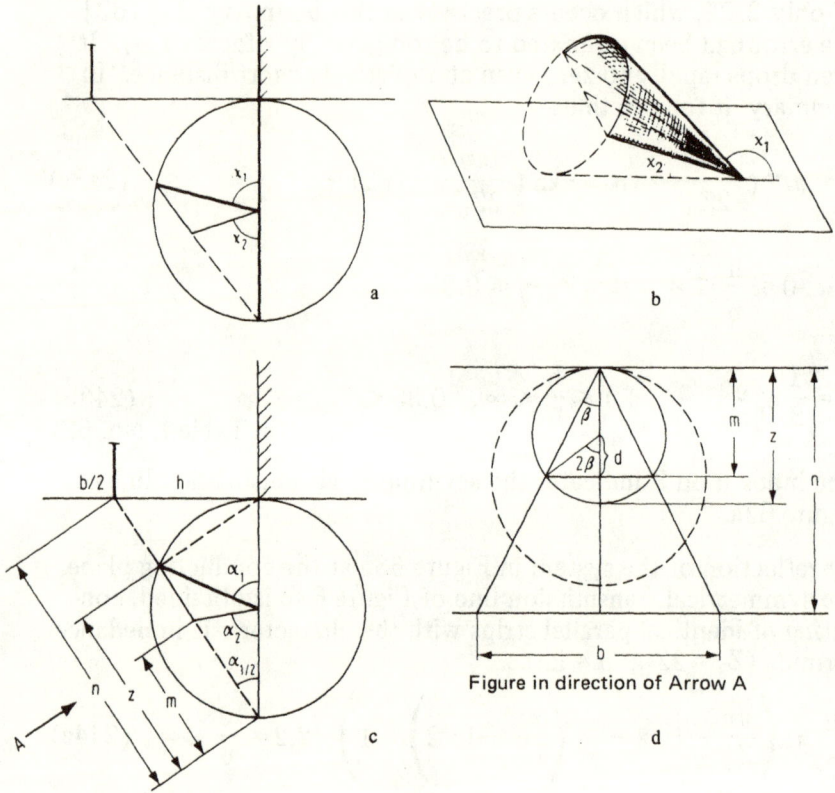

Figure 64

The desired projection equation then follows, in the form:

$$\frac{b}{h} = \frac{2\sqrt{1+(\tan \alpha_1/2)^2}}{\tan \alpha_1/2} \tan \left[\text{arc } \cos \left(\frac{\sin \alpha_2}{2 \sin(\alpha_2 + \alpha_1/2) \cos \alpha_1/2} \right)^{1/2} \right]. \tag{246}$$

The voluminous expression obtained after substitution in (243) will not be written down here. Also, we would like to discuss briefly below the projections now possible of the conical transmission line onto the plane.

Clearly, the hollow cone segment in the projection of Figure 65a provides the transmission line of 65b. Furthermore, the parallel lines of the associated Figures 66b, 67b, 68b, 70b, are formed by the projections according to Figures 66a, 67a, 68a, 70a. Finally,

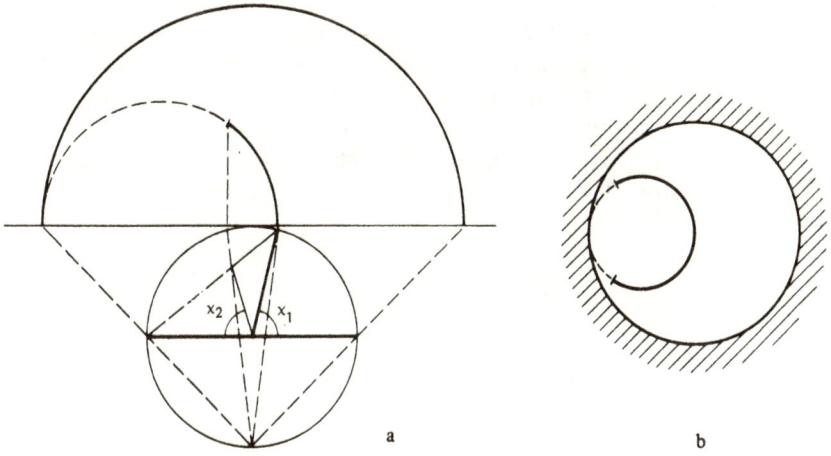

Figure 65

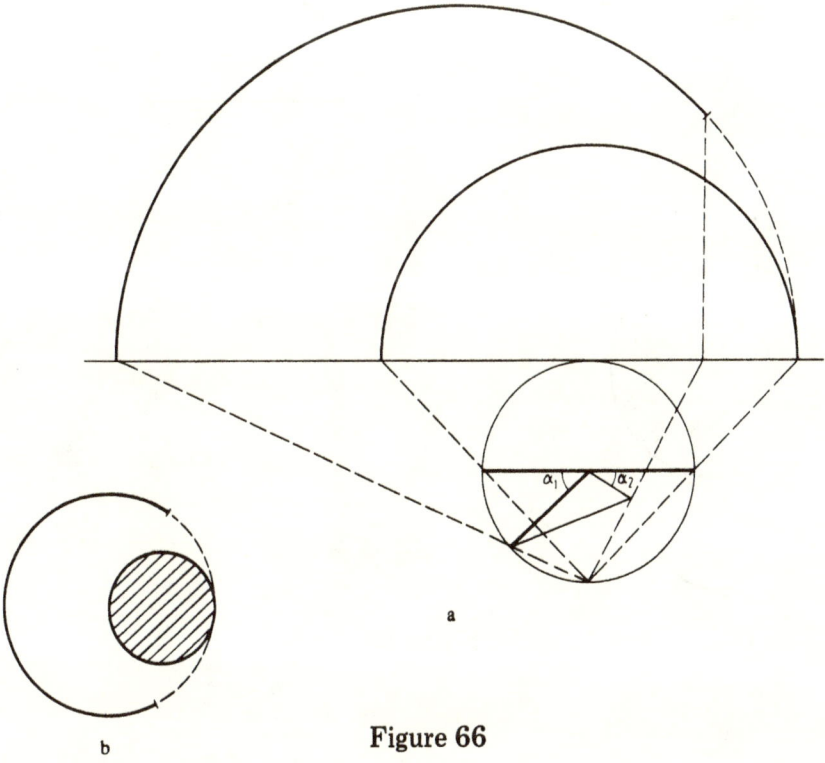

Figure 66

six different types of parallel transmission lines are associated here with a single conical line. (With respect to a practical calculation of these parallel transmission lines, however, it is better here not to take the route through the conical line with its voluminous formula, but rather to carry out some transformations $w - 1/(z - z_0)$, with the positions of the reference points outlined in Figure 71.)

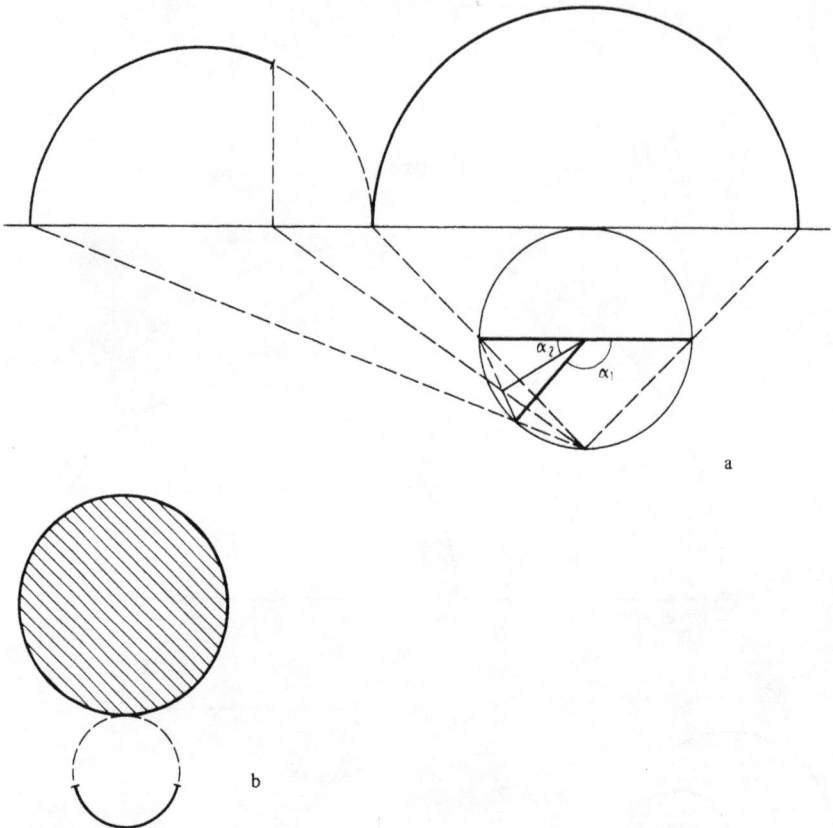

Figure 67

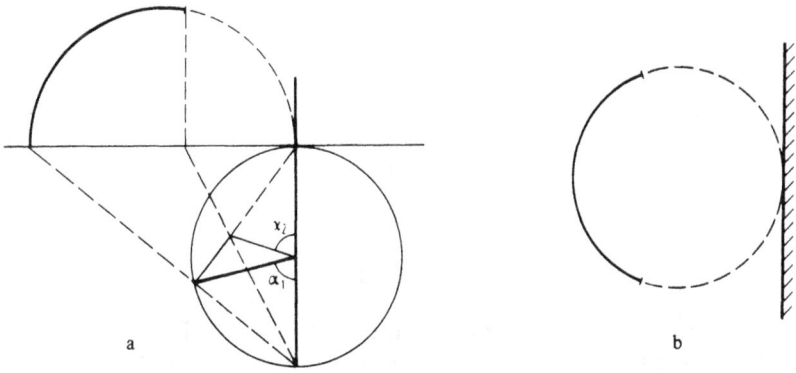

Figure 68

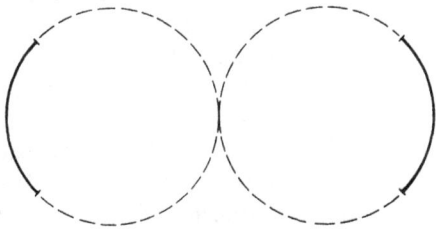

Figure 69

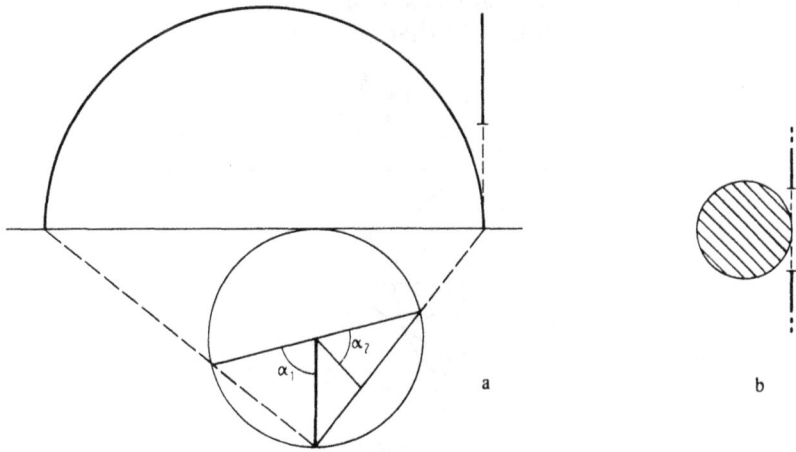

Figure 70

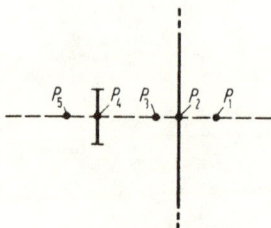

Figure 71

In conclusion, let us consider a few more transformation possibilities of the transmission line of Figure 63a. We can start from the cylinder opposed to the edge of a conductive plane in Figure 53b (with an arbitrary angle γ). From this, as shown, is obtained the circular cone opposed to the edge of a plane, e.g., as in Figure 55a. If this system is projected in the position of Figure 72 onto the plane, the result is precisely the strip parallel to a conductive plane as in Figure 63a. Also, the systems treated earlier in Figures 55b, 56b, and 57b, are then clearly related. If in addition, the cone in Figure 72 is swiveled somewhat in the clockwise direction, the system of Figure 73 can also be obtained by projection, as an example. This can again be converted into a conical transmission line, as in Figure 74, which is identical with the inclined conical strip on a conductive plane as in Figure 43c. Various extensive transformation series connected with one another can therefore be formulated. For this reason, it would undoubtedly be useful to calculate one of the conical transmission lines, e.g., from Figure 43a or Figure 54, more precisely than was done in the course of this work.

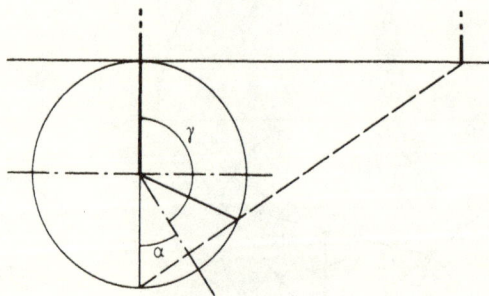

Figure 72

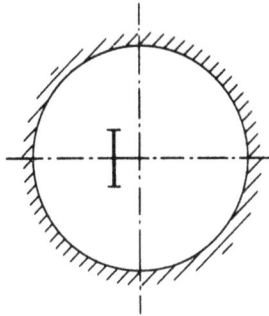

Figure 73

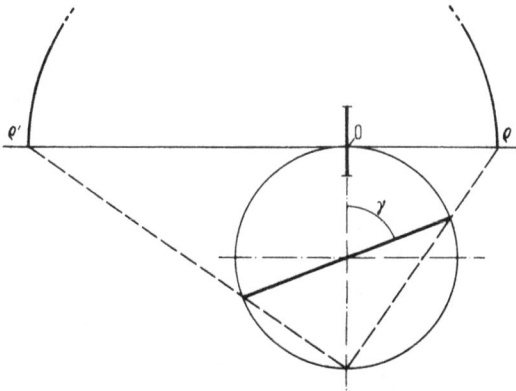

Figure 74

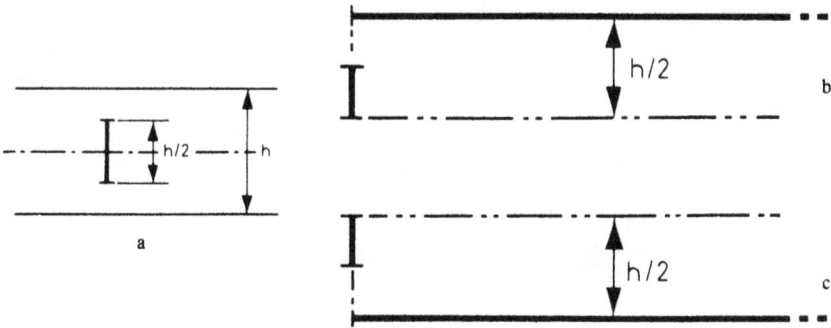

Figure 75

Chapter VIII

Supplements

1. Use of Complementarity

Both in the reference system of the double conical strips and also in many other systems, complementarity considerations can be employed with advantage, in order to obtain directly an exact value in the center of a parameter region. Let us consider first a case already well-known, namely the transverse strip of Figure 40a, specifically in a symmetrical position. For reasons of symmetry, it suffices to consider one-fourth of the cross section of Figure 75a with the electrical and magnetic walls sketched there, as in Figure 75b. The interchange of the walls leads to Figure 75c. If the vertical electrical section of the wall is now just as long as the magnetic section of wall, nothing changes after all with respect to the characteristic impedance from complementation, i.e., because of (14)b, $Z_m = \eta$, and for the particular system in Figure 75a, the characteristic impedance is finally obtained as $Z = \eta/4$.

As the next example, we might consider the shielded transmission line in Figure 41a. If the electrical walls here are replaced by magnetic walls, and vice versa, the system of Figure 76a is obtained. Since the two layers are apparently completely symmetrical, and can be added to upward and downward to any arbitrary extent, a section can also be considered to be isolated as in Figure 76b. However, this system corresponds to a very good approximation to an electrical plate with large ϵ, coated on both sides with a conductive layer, with the exception of two slits. The complement to (176) then applies:

$$k_A = e^{\pi u/h} \ , \quad \lambda = 2 \ . \tag{247}$$

(in Table I, No. 64)

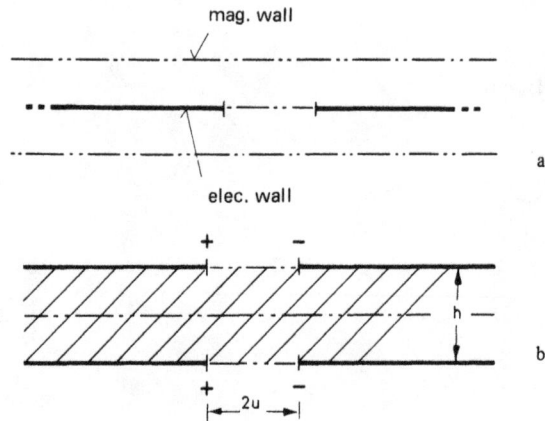

Figure 76

The capacity, which in this case is probably of more frequent interest in practice, is obtained from this with (2).

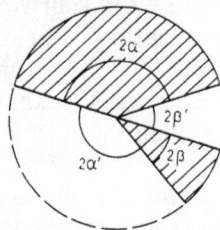

Figure 77

Such considerations of symmetrical complementary systems naturally provide particularly great advantages wherever the general characteristic impedance formulas are very extensive, such as in the case of various large planar and non-coaxial conical strips, for example. If it is desired here to provide equivalent systems by the interchange of electrical and magnetic surfaces, as in Figure 77, then the following conditions apply:

$$2\alpha + 2\beta' = \pi, \quad \alpha = \alpha', \quad \beta = \beta'. \tag{248}$$

If these conditions are met for the case drawn, then the characteristic impedance is exactly $Z_m = \eta/2$.

2. Conversions

Other equations than those derived here are often found in books
and tables. Conversions are therefore frequently needed for comparison. For this reason, the most important conversion equations
might be summarized briefly.

First are the exact equations

$$\ln x = \text{ar cosh } \frac{1}{2}\left(x + \frac{1}{x}\right), \tag{249}$$

$$\text{ar cosh } x = \ln(x \pm \sqrt{x^2 - 1}), \tag{250}$$

$$2 \text{ ar cosh } x = \text{ar cosh}(2x^2 - 1). \tag{251}$$

Then, however, the following approximations are often used for
large x:

$$\ln x \approx \text{ar cosh } x/2 \tag{252}$$

or

$$\text{ar cosh } x \approx \ln 2x .$$

3. Field Spaces With Layered Dielectric Media (ϵ_{eff})

The calculations up to this point have been concerned with transmission lines with arbitrarily good metal surface conductors and
with a field space which is filled with air or a loss-free homogeneous
medium. However, it is also relatively frequently the case that the
field space is filled with a non-homogeneous medium consisting of
a layered dielectric. In this case, as is well-known, no TEM wave
can be formed. However, there are numerous relationships which
can still be approximated with the concepts of the TEM transmission line waves. In this connection, it is desirable to work with
the so-called effective dielectric constant. As a rule, it is determined by measurement, for example, as in [24]. With Z_0 as the
characteristic impedance for the transmission line in air, the characteristic impedance for the line with dielectric layers is then:

$$Z = Z_0/\sqrt{\epsilon_{eff}} . \tag{253}$$

The limitation

$$1 \leqslant \epsilon_{\text{eff}} \leqslant \epsilon_r \qquad\qquad (254)$$

obviously applies in this case.

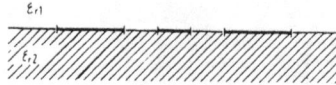

Figure 78

In certain cases, useful approximations for ϵ_{eff} can be found also without measurement or tedious calculation. This is true, for example, for transmission lines with symmetrical dielectric coatings, if it is desired to determine the input impedance of an arbitrarily long or terminated transmission line, or if dispersion phenomena can still be neglected. Let us take as an example the stripline with the cross section of Figure 78. Let a voltage jump be suddenly applied to the input of such a line. Then, to a first approximation, if each half-space is considered in its own right, the voltage waves would be propagated along the transmission lines (perpendicular to the plane of the sheet) at different velocities, at a greater velocity in the upper half-space (e.g., air with $\epsilon_{r_1} = 1$) than in the lower half-space (dielectric with $\epsilon_{r_2} > 1$). However, the faster moving portion will increasingly overlap into the lower space, i.e., viewed as a whole, a flattening of the pulse edges, or a dispersion, takes place. However, this is not important for the input impedance of such an infinitely long transmission line, which can be composed of the impedances of the isolated half-spaces (the plane of partition between the two spaces then consists of electrical and magnetic walls, to a first approximation). Also, for short line sections, the dispersion does not have remarkable effect on the characteristic impedance. The equations introduced above produce the characteristic impedance $2\,Z_0/\sqrt{\epsilon_{r1}}$ for the upper half-space, and the impedance $2\,Z_0/\sqrt{\epsilon_{r2}}$ for the lower. The parallel circuit then has the input impedance Z_i:

$$Z_i = \frac{2}{\sqrt{\epsilon_{r1}} + \sqrt{\epsilon_{r2}}}\,Z_0\,, \quad \sqrt{\epsilon_{\text{eff}}} = \frac{\sqrt{\epsilon_{r1}} + \sqrt{\epsilon_{r2}}}{2}. \qquad (255)$$

(in Table 1, No. 65)

If characteristic impedances of coplanar transmission lines are com-
pared, which have been calculated on the one hand in accordance
with (255) and measured on short sections on the other hand (as
in [16] for example, where curve a_4 must be multiplied by 4/3),
an error is found of the order of magnitude of approximately 15%.
Considering the factors neglected, this seems to be surprisingly
good.

In cases in which relatively thin dielectric plates are used as carriers
for metallic strips, ϵ_{eff} can no longer be determined so easily by
computation. The methods used here cannot be illustrated further
within the framework of this paper, however. Also, it is very
difficult in addition to calculate the variation of ϵ_{eff} with frequency.

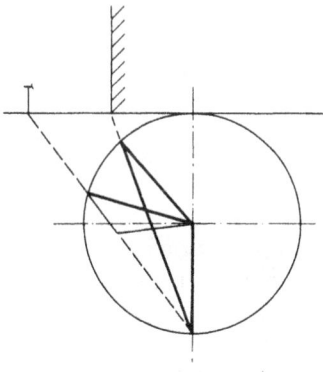
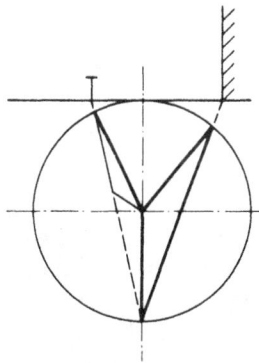

Figure 79   Figure 80

4. Final Consideration of the Stereographic Relationships Between
Conical and Cylindrical Transmission Lines

If we consider the point of contact of the unit sphere with the
plane to be fixed, a number of conical transmission lines of the
same characteristic impedance can be generated by shifting the
cross sectional image of a cylindrical line in the plane and by subsequent stereographic projection. Let us illustrate this once more
by an example, with the use of Figures 64a, 79, 80, 81, and 82.
The cylindrical line of Figure 63a has been projected here onto a
sphere in various positions. Among the conical transmission lines
generated there are some of a general type in which are contained

all other conical transmission lines by special choice of parameters. This is the case in Figures 79, 80, and 82, where the axial directions and aperture angles of the two cones, as well as the size of the hollow cone segment, can be chosen arbitrarily. The conical transmission lines of Figures 64a and 81, on the other hand, are much more specific, and the others cannot be derived from them.

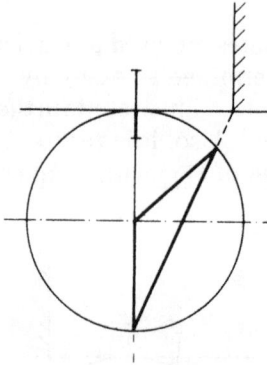

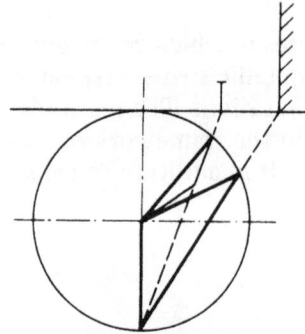

Figure 81   Figure 82

Besides such examples, there are also others in which a conical transmission line of a general type can apparently not be obtained by translations of the cross section into the plane with subsequent stereographic projection, even if the conical line generated is rotated again in the plane, and then is portrayed once more on the sphere. This had become apparent, for example, at the very beginning in the group of circular conical transmission lines and circular cylindrical lines. Translations in the plane, rotation of the sphere, and stereographic projection are therefore not sufficient in general for the production of the conical line of the more general type. This is probably because none of the operations mentioned is equivalent to a complete linear transformation. We would like to consider this with the example of the rotation of the cylinder with subsequent projection.

If we rotate the sphere in Figure 83 by the angle ϑ, the point P_1 passes over to point P_2, and the corresponding point ρ_1 in the plane changes to ρ_2. With the projection equations

$$\rho_1 = \tan \vartheta_1/2, \quad \rho_2 = \tan \vartheta_2/2, \tag{256a}$$

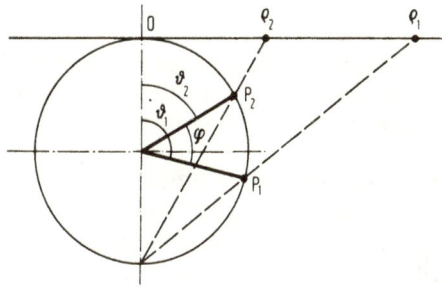

Figure 83

and $\vartheta_1 - \vartheta_2 = \varphi$, ρ_2 can now be written as a function of ρ_1 and φ, and there is obtained:

$$\rho_2 = \tan\left(\text{arc}\tan\rho_1 - \frac{\varphi}{2}\right). \qquad (256b)$$

This can be transformed with the addition theorem for $\tan(\alpha-\beta)$ to give:

$$\rho_2 = \frac{-\tan\dfrac{\varphi}{2} + \rho_1}{1 + \tan\dfrac{\varphi}{2} \cdot \rho_1} \qquad (256c)$$

As a test, we can set $\varphi = \pi$, i.e., we can bring point P_2 into the position opposite P_1, for which the relationship $\rho_2 = -1/\rho_1$ is then produced (cf. (20)).

Since we had initially limited ourselves to the consideration of real parameters, for real z, (256c) should be identical with a linear transformation

$$w = \frac{a+bz}{c+dz}, \qquad (257)$$

or a circular relationship of the second type [31]. Let us now set aside the differentiation of circular relationships of the first and second kind as unimportant for our technical application, so that the coefficients of a transition to real z must remain un-affected. By comparison of (256c) with (257), it can now be recognized that the coefficients a, b, c, d cannot be chosen arbi-trarily, but that specifically, a = −d, b = c. Similar limitations can also be stated for all other component operations. Only the combination of translation and change of scale in the plane, stereographic projection, and rotation of the sphere produces all conical transmission lines, particularly of the more general type.

Finally, however, if a single more general conical transmission line is selected, all cylindrical transmission lines can be obtained by simple rotation and projection. For this reason, the statement is probably justified that suitably selected conical transmission lines have a more general character, since they include both the entire group of cylindrical lines related linearly to one another, as well as the corresponding conical lines.

Furthermore, if as in many of the examples shown, a reflection is executed in addition to the stereographic projection, this operation in general leads outside of the scope of linear relationships, and therefore also produces a conical transmission line of a new type. However, if one of the conical conductors is a circular cone, no expansion is produced. Reflection is therefore essentially a non-linear operation. In the generation of complementary transmission lines, some discrimination must also be applied. In this paper, only examples were needed in which the complementary line is identical in form with the initial line. For this reason, we obtain no expansion of the already known group of lines in this way. The advantage of its utilization was principally in the use of approximation equations. In general, however, the form of the complementary transmission line does not have to be identical with the initially presented line, as can be seen in Figure 4f, g, for example, so that with its help, the group of linear related transmission lines can also be broadened.

5. *Derivation of the Complementarity Theorem (Babinet's Principle)*

If one starts with the Maxwell Equations in the form:

$$\text{rot } H = i\omega\epsilon E$$
$$\text{rot } E = -i\omega\mu H, \tag{258}$$

and uses them in the field space of a transmission line symmetrical to a plane as outlined in the example of Figure 84a, then as is well-known, a configuration of E and H as indicated there in two lines,

satisfies these equations. E is perpendicular to the conductor surface, E and H are perpendicular to one another, and H is perpendicular to the plane of symmetry. New parameters E' and H' can
now be defined as follows:

$$E' = -\sqrt{\frac{\mu}{\epsilon}}\, H$$

$$H' = \sqrt{\frac{\epsilon}{\mu}}\, E\,.$$

(259)

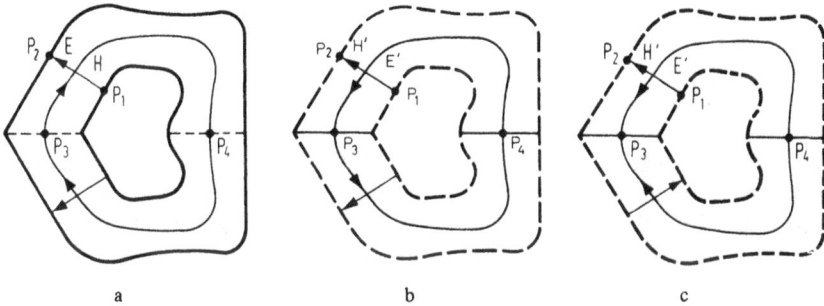

Figure 84

If the equations are turned around, they read:

$$H = -\sqrt{\frac{\epsilon}{\mu}}\, E'$$

$$E = \sqrt{\frac{\mu}{\epsilon}}\, H'\,.$$

(260)

If these parameters are substituted in the Maxwell Equations, it
follows that:

$$\text{rot } E' = -i\omega\mu H'$$

$$\text{rot } H' = i\omega\epsilon E'\,.$$

(261)

In such a conversion, therefore, the new parameters also satisfy
the Maxwell Equations. (The wave formed from E' and H' merely
has the opposite direction of propagation because of the reversal

of the sign in one parameter.) If the boundary surfaces with arbitrarily high electrical conductivity ($\kappa = \infty$) are now interchanged with those of arbitrarily high magnetic susceptibility ($\mu = \infty$) and vice versa, the shape of the field "lines" remains completely unchanged. This is outlined in Figure 84b.

One might next consider only a portion of the field space enclosed by the so-called electrical and magnetic walls on the side of the plane of symmetry, as in Figure 84a. Let us call the characteristic impedance of this Z_{T_1}, which is obtained from

$$Z_{T_1} = \frac{\int_1^2 \text{Eds}}{\int_3^4 \text{Hdl}} . \tag{262}$$

For better differentiation, the first path from 1 to 2 is divided into path elements ds and the second path from 3 to 4 into path elements dl. The characteristic impedance Z_{T_2} of the complementary partial space in Figure 84b then correspondingly is:

$$Z_{T_2} = \frac{\int_3^4 \text{E'dl}}{\int_1^2 \text{H'ds}} . \tag{263}$$

If the parameters E' and H' from equation (259) are now introduced into the last equation, and the negative sign is suppressed, then it follows that:

$$Z_{T_2} = \frac{\sqrt{\mu/\epsilon}\,\int \text{Hdl}}{\sqrt{\epsilon/\mu}\,\int \text{Eds}} = \frac{\mu}{\epsilon}\,\frac{\int \text{Hdl}}{\int \text{Eds}} . \tag{264}$$

If equation (262) is multiplied by equation (264) we find further that:

$$Z_{T_1} \cdot Z_{T_2} = \frac{\mu}{\epsilon} = \eta^2 . \tag{265}$$

If it is desired to formulate the corresponding equations for the entire system consisting of two partial spaces, then an additional small consideration must be appended. It is clear initially that with E' and H', the negative parameters —E' and —H' also satisfy the Maxwell Equations. The two partial spaces can therefore be combined as in Figure 84b or 84c to form an overall system. However, the latter possibility appears primarily suitable physically.

The characteristic impedance Z_I of the entire transmission line of Figure 84a is given by:

$$Z_I = \frac{\oint Eds}{\oint Hdl} = \frac{\oint Eds}{2 \oint Hdl} \ . \tag{266}$$

The impedance Z_{II} of the entire complementary transmission line in Figure 84c is:

$$Z_{II} = \frac{\oint E'dl}{\oint H'ds} = \frac{\oint E'dl}{2 \oint H'ds} \ . \tag{267}$$

If the parameters H' and E' from Equation (259) are introduced into the latter equation, and if Equation (266) is multiplied by Equation (267), then finally there is obtained:

$$Z_I \cdot Z_{II} = \frac{\mu}{4\epsilon} = \frac{\eta^2}{4} \ . \tag{268}$$

(Equation (268) was first given by Booker [25] for slot antennas. The same equation was derived for cylindrical transmission lines in [17], wherein (265) also resulted for lines with simply related boundaries.)

The result (268) is also obtained by observing that as a result of the parallel connection of two identical partial spaces in Figure 84c, the following is valid:

$$Z_I = Z_{T_1}/2 \ ; \quad Z_{II} = Z_{T_2}/2 \ . \tag{269}$$

Now walls with good magnetic susceptibility can be made physically in substantially poorer approximation than walls with good electrical conductivity. For this reason, such considerations can be used to great advantage generally only in those special transmission lines in which no real magnetic walls are necessary because of the complementation, such as for parallel striplines in Figure 4a and Figure 4b, for example, and for conical striplines in Figure 4c and Figure 4d. That is to say, there are no magnetic bodies here, but only planes penetrated vertically by the H-lines, i.e., the introduction of real magnetic walls here has no effect and they can therefore be omitted.

Another important equation is obtained for the case $Z_I = Z_{II}$. This average value was designated as Z_m in (77). From Equation (268), it follows that:

$$Z_m = \eta/2 .$$
(270)

As particularly clear examples, a coplanar parallel stripline with $r'/r = 2$ is shown in Figure 85a, and a double conical stripline with $\alpha = \pi/4$ in Figure 85b, which have the characteristic impedance $Z_m = \eta$.

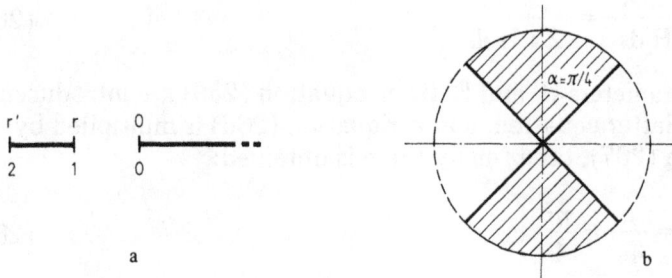

Figure 85

6. Overcoupling and Decoupling Between Several Lines

As already stated in the introduction, fundamentally all transmission line calculations carried out here can also be undertaken without the use of the stereographic projection. It merely facilitates the calculation and the comprehensibility. This is true above all for the systems of more than 2 lines. It is well known that the coupling phenomena are of interest here, such as unintentional overcoupling (cross-talk), intentional overcoupling (directional coupling) and the conditions for complete decoupling. The first two points were discussed in great detail in [21] and [15], while the latter was treated in [22]. However, since so much use has been made of stereographic projection in the present paper, it will be described in conclusion with the example of decoupling, how it can serve not only for the calculation of the characteristic impedance of transmission lines, but also for the formulation of more general situations.

It has already been long known that a maximum of three lines can be exactly decoupled from one another. Breisig [26] illustrated this for vanishingly thin conductors in the following way; refer to the cross section of the lines in Figure 86. Equipotential surfaces can be determined for a first pair of conductors I I'. If a second pair of conductors II II' is constructed in such a way that the two conductors lie on such a potential surface, then the two transmission lines are exactly decoupled from one another. Equipotential surfaces can now be determined for the second pair of conductors, and the conductors III III' of the third conductor can be accommodated on them at the points of intersection with the equipotential surface used previously. Since the conductors therefore lie on equipotential surfaces of the first and second lines, they are exactly decoupled from both of these transmission lines. Finally, it therefore follows that each line is decoupled from every other.

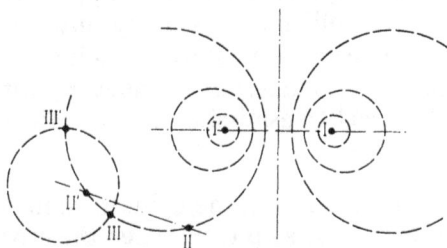

Figure 86

In this way, the decoupling of very thin conductors can be quite well described with the concept of equipotential surfaces. However, if it is desired to express the decoupling conditions purely geometrically, it becomes more involved (although it is still not very difficult for the vanishingly thin conductors). The transition to conical transition lines, however, leads immediately to a simple geometrical description.

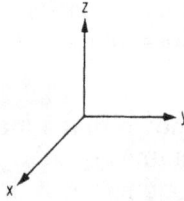

Figure 87

If we choose the origin of the coordinate system in Figure 87 as
the feed point of the conical transmission lines, the line I, for
example, can lie in such a way that the conductor I coincides
with the positive y-axis and the conductor I' with the negative
y-axis. The conductors of the second transmission line must now
lie on an equipotential surface, which here are concentric circular
cones around the positive or negative y-axis. We now require
that the conductors II and II' be symmetrical to the y-z plane.
Since this coordinate system is arbitrary, and can also be rotated
around the y-axis, for example, this symmetry condition is
necessary and sufficient for exact decoupling. The third transmission line can now be placed symmetrically to the x-y plane
and in the y-z plane for decoupling, since two equipotential
surfaces of the first two transmission lines always intersect here.
By projection of the three decoupled conical lines into various
angular positions on the plane, numerous conductor positions
of three decoupled cylindrical transmission lines can finally be
obtained.

The advantage of the conversion from the cylindrical lines to the
conical lines, however, only shows up truly when the cross sections
of the conductors are no longer subjected to restrictions, i.e., when
we consider thick conductors or conductors consisting of wide
strips. In these cases, it is generally difficult to work with equipotential planes, and it is advisable to refer to the concepts of odd
and even modes for the discussion of the decoupling. It then turns
out [22] that the decoupled non-transmission lines must satisfy
more stringent symmetry requirements than the thin ones.

If we again start from the coordinate system in Figure 87 and let
the axis of the first line coincide with the y-axis, then it follows
from simple considerations that the axes of the second line lie in
the x-y plane and symmetrically to the y-z plane, while the axes

of the third line must then lie in the y-z plane and symmetrically to the x-y plane. Nothing else is assumed here concerning the shape of the conical conductors. By stereographic projection, one again obtains all possibilities for the placement of exactly decoupled cylindrical transmission lines.

It can be demonstrated that the decoupling conditions in the plane would only be extraordinarily tedious to formulate, while in three dimensions, probably no simpler symmetry criteria can be found.

Chapter IX

References for Part A

[1]   Hilberg, W., "On the Possibility of Replacing Certain Characteristic Impedance Formulas Which Contain Elliptical Integrals by Approximation Formulas of Arbitrarily Selectable Precision," *AEU 21* (1967), No. 11, pp. 603-616.

[2]   Weber, E., *Electromagnetic Fields, Theory and Applications, Volume I — Mapping of Fields*, Wiley-Verlag, New York, London (1950).

[3]   Schelkunoff, S.A., *Advanced Antenna Theory*, Wiley-Verlag, London (1952).

[4]   Rothe, P.G., "Approximate Formulas for the Characteristic Impedances of Some Conical Transmission Lines," *Royal Aircraft Establishment, Farnborough, Hants, England Technical Note No. RAD. 532*, 1953.

[5]   Küpfmüller, K., *Introduction to Theoretical Electrical Engineering*, Springer-Verlag.

[6]   Smythe, W.R., *Static and Dynamic Electricity*, McGraw-Hill-Verlag, New York, 1950.

[7]   Collin, R.E., *Field Theory of Guided Waves*, McGraw-Hill-Verlag, New York, 1960.

[8]   Becker-Sauter, *Theory of Electricity*, Teubner-Verlag, Stuttgart, 1957.

[9]   Buchholz, H., *Electrical and Magnetic Potential Fields*, Springer-Verlag, 1957.

[10]   Mangold and Knopp, *Introduction to Higher Mathematics, Volume II*, Hirzel-Verlag, 1958.

[11]   Oberhettinger, F., and W. Magnus, *Application of Elliptical Functions to Physics and Engineering*, Springer-Verlag, 1959.

[12]   Magnus, W., and F. Oberhettinger, "The Calculation of the Characteristic Impedance of a Stripline with Circular or Rectangular Cross Section of the Outer Conductor," *Archiv fur Elektrotechnik*, Vol. 37 (1943), No. 8, pp. 380-390.

[13]   Meinke − Gundlach, *Handbook of High Frequency Engineering*, Springer-Verlag.

[14]   Megla, G., *Decimeter Wave Engineering*, Berliner Union Stuttgart.

[15]   Hilberg, W., "Possibilities and Limitations of an Elementary Theory of Overcoupling of Pulses and Sine Waves Between Parallel Transmission Lines," *NTZ-Report*, No. 4, VDE-Verlag, Berlin, 1969.

[16]   Kilkowski, J., "The Characteristic Impedance of a Stripline Prepared with the Use of the Foil Etching Process," *Telefunken*, 1964.

[17]   Hilberg, W., "A Simple and Good Approximation for the Characteristic Impedance of Parallel Striplines," *AEU 22* (1968), No. 3, pp. 122-126, addendum by W. Nowak in *AEU 23* (1969), No. 8, pp. 430.

[18]   Booker, H.G., "Slot Aerials and Their Relation to Complementary Wire Aerials (Babinet's Principle)," *The Journal of IEE*, Vol. 93, Part III A, No. 4, 1946, pp. 626-629.

[19]   Hilberg, W., "From Approximations to Exact Relations for Characteristic Impedances," *IEEE Transactions MTT*, Vol. 17, No. 5, May 1969, pp. 259-265.

[20]   Knopp, K., "Elements of Function Theory," "Function Theory I, II," "Compilation of Projects on Function Theory," *Goshen Compilation*, Vol. 1109, 668, 703, 887.

[21]   Hilberg, W., "Elementary Treatment of Overcoupling of
        Pulses and Sine Waves Between Parallel Transmission
        Lines," *NTZ*, June 1969, No. 6, pp. 368-373.

[22]   Hilberg, W., "The Decoupling of Lines Consisting of Two
        Conductors Each," *AEU 22* (1968), No. 1, pp. 39-45.

[23]   Hilberg, W., "Stringent Calculation of the Characteristic
        Impedance of Parallel Striplines and a Comparison with
        Approximations," *AFE 54* (1971), pp. 200-205.

[24]   Deutsch, J., and H.J. Jung, "Measurement of the Effective
        Dielectric Constant of Microstrip Transmission Lines in the
        Frequency Range 2 GHz to 12 GHz," *NTZ*, 1970, No. 12,
        pp. 620-624.

[25]   Booker, H.G., "Slot Aerials and Their Relation to Com-
        plementary Wire Aerials (Babinet's Principle)," *The
        Journal of IEE*, Vol. 93, Part III A, No. 4, 1946, pp. 620-
        626.

[26]   Breisig, F., *Theoretical Telegraphy*, Verlag Vieweg,
        Braunschweig 1910.

[27]   Johari, O., and G. Thomas, *The Stereographic Projection
        and Its Applications*, Interscience Publishers (Wiley),
        New York, 1969.

[28]   Hilberg, W., "Approximations for the Elliptical Integral
        Functions K/K' and Recursions for the Optional Improve-
        ment of Its Accuracy, Particularly for the Calculation of
        Characteristic Impedance," *Archiv fur Elektrotechnik*, Vol.
        53, 1970, No. 5, pp. 290-298.

[29]   Koppenfels, W.V. and F. Stallmann, *Conformal Mapping in
        Practice*, Springer-Verlag, 1959.

[30]   Kaden, H., *Eddy Currents and Shielding in Information
        Technology*, Springer-Verlag, 1959.

[31]   Caratheodory, C., *Theory of Functions I*, Birkhauser
        Verlag Basel, 1950.

Part B

Comments: The relative error of the approximations in which the parameters Z_h, Z_1, and Z_m are indicated, is always smaller than $2.4 \cdot 10^{-3}$. See Equation (77). The boundary between range of high impedance values Z_h and the range for low impedance values Z_1 is given by Z_m. The constant η is defined in Equation (15).

Cons. no.	System	Formulas	Equation & Dia. no.
1		$$Z = \frac{\eta}{2\pi} \ln r'/r$$	33 D1
2		$$Z = \frac{\eta}{2\pi}\left(\text{arcosh }\frac{A_1}{a} - \text{arcosh }\frac{A_2}{a}\right)$$	29 D2
3		$$Z = \frac{\eta}{2\pi}\text{ arcosh }\frac{D^2+d^2-4e^2}{2Dd}$$	48 D3

Cons. no.	System	Formulas	Equation & Dia. no.
4		$$Z = \frac{\eta}{2\pi} \; \text{arcosh} \; \frac{4D^2 - d_1^2 - d_2^2}{2\,d_1\,d_2}$$	56 D4
5		$$Z = \frac{\eta}{\pi} \; \text{arcosh} \; \frac{D}{d}$$	59 D5
6		$$Z = \frac{\eta}{2\pi} \; \text{arcosh} \; \frac{2L}{d}$$ $$Z = \frac{\eta}{2\pi} \; \text{arcosh} \; \frac{\sin \gamma}{\sin \beta}$$	60 199 D6

Cons. no.	System	Formulas	Equation & Dia. no.
7		$Z = \dfrac{\eta}{2\pi}\operatorname{arcosh}\dfrac{A_1}{a}$; $a^2 = A_1^2 - B_1^2$	29 D6
8		$Z_h = \dfrac{\eta}{2\pi}\ln\left[2\,\dfrac{r'}{r}\right]$ $Z_l = \dfrac{\pi\eta}{8}\Big/\ln\left[2\,\dfrac{r'+r}{r'-r}\right]$ $Z_m = \eta/4$	37 D7
9		$Z_h = \dfrac{\eta}{\pi}\ln\left[2\,\dfrac{\sqrt{D'/D}+1}{\sqrt{D'/D}-1}\right]$ $Z_l = \dfrac{\pi\eta}{4}\Big/\ln\left[2\sqrt{D'/D}\right]$ $Z_m = \eta/2$	93 D8

Cons. no.	System	Formulas	Equation & Dia. no.
10		$Z_h = \dfrac{\eta}{\pi} \ln\left[2\cot\alpha/2\right]; \ \infty \geq Z_h \geq Z_m; \ \dfrac{\pi}{4} \geq \alpha \geq 0; \ \dfrac{\Delta Z}{Z} \leq 0.237\%$ $Z_l = \dfrac{\pi\eta}{4} \Big/ \ln\left[2\cot\left(\dfrac{\pi}{4} - \dfrac{\alpha}{2}\right)\right]; \ Z_m \geq Z_l \geq 0; \ \dfrac{\pi}{2} \geq \alpha \geq \dfrac{\pi}{4}; \ \dfrac{\Delta Z}{Z} \leq 0.236\%$ $Z_m = \eta/2; \ \alpha_m = \pi/4.$	75 76 9 D9
11		$Z_h = \dfrac{\eta}{2\pi} \ln\left[2\dfrac{\sqrt{r'/r}+1}{\sqrt{r'/r}-1}\right]$ $Z_l = \dfrac{\pi\eta}{8} \Big/ \ln\left[2\sqrt{r'/r}\right]$ $Z_m = \eta/4$	92 D10
12		$Z_h = \dfrac{\eta}{2\pi} \ln\left[2\cot\alpha/2\right]; \ \dfrac{\pi}{4} \geq \alpha \geq 0$ $Z_l = \dfrac{\pi\eta}{8} \Big/ \ln\left[2\cot\left(\dfrac{\pi}{4} - \dfrac{\alpha}{2}\right)\right]; \ \dfrac{\pi}{2} \geq \alpha \geq \dfrac{\pi}{4}$ $Z_m = \eta/4; \ \alpha_m = \pi/4$	74 85 D11

Cons. no.	System	Formulas	Equation & Dia. no.
13		$Z_h = \dfrac{\eta}{2\pi} \ln\left[2\sqrt{\dfrac{(r-a)(r'+a)/(r+a)(r'-a)+1}{(r-a)(r'+a)/(r+a)(r'-a)-1}}\right]$ $Z_l = \dfrac{\pi\eta}{8}/\ln\left[2\sqrt{(r-a)(r'+a)/(r+a)(r'-a)}\right]$ $Z_m = \eta/4$	107 D 12
14		$Z_h = \dfrac{\eta}{2\pi} \ln\left[2\sqrt{\dfrac{(a-r)(a+r')/(a+r)(a-r')+1}{(a-r)(a+r')/(a+r)(a-r')-1}}\right]$ $Z_l = \dfrac{\pi\eta}{8}/\ln\left[2\sqrt{(a-r)(a+r')/(a+r)(a-r')}\right]$ $Z_m = \eta/4$	105 D 13
15		$Z_h = \dfrac{\eta}{2\pi} \ln\left[2\sqrt{\dfrac{\tan\frac{\gamma+\alpha}{2}/\tan\frac{\gamma-\alpha}{2}+1}{\tan\frac{\gamma+\alpha}{2}/\tan\frac{\gamma-\alpha}{2}-1}}\right]$ $Z_l = \dfrac{\pi\eta}{8}/\ln\left[2\sqrt{\tan\frac{\gamma+\alpha}{2}/\tan\frac{\gamma-\alpha}{2}}\right]$ $Z_m = \eta/4$	100 D 14

Cons. no.	System	Formulas	Equation & Dia. no.
16		$$Z_h = \frac{\eta}{\pi} \ln\left[2\,\frac{\sqrt{\tan\frac{\gamma+\alpha}{2}/\tan\frac{\gamma-\alpha}{2}}+1}{\sqrt{\tan\frac{\gamma+\alpha}{2}/\tan\frac{\gamma-\alpha}{2}}-1}\right]$$ $$Z_l = \frac{\pi\eta}{4}\Big/\ln\left[2\sqrt{\tan\frac{\gamma+\alpha}{2}/\tan\frac{\gamma-\alpha}{2}}\right]$$ $$Z_m = \eta/2$$	101 D 15
17		$$Z_h = \frac{\eta}{2\pi} \ln\left[2\,\frac{\sqrt{r'/r}+1}{\sqrt{r'/r}-1}\right]$$ $$Z_l = \frac{\pi\eta}{8}\Big/\ln\left[2\sqrt{r'/r}\right]$$ $$Z_m = \eta/4$$	97 D 10
18		$$Z_h = \frac{\eta}{2\pi} \ln\left[2\,\frac{\sqrt{\tan\frac{\gamma+\alpha}{2}/\tan\frac{\gamma-\alpha}{2}}+1}{\sqrt{\tan\frac{\gamma+\alpha}{2}/\tan\frac{\gamma-\alpha}{2}}-1}\right]$$ $$Z_l = \frac{\pi\eta}{8}\Big/\ln\left[2\sqrt{\tan\frac{\gamma+\alpha}{2}/\tan\frac{\gamma-\alpha}{2}}\right]$$ $$Z_m = \eta/4$$	109 D 14

Cons. no.	System	Formulas	Equation & Dia. no.
19		$Z_h = \dfrac{\eta}{2\pi}\ \ln\left[2\sqrt{\dfrac{\tan\dfrac{\pi}{2}\dfrac{\gamma+\alpha}{\vartheta}\Big/\tan\dfrac{\pi}{2}\dfrac{\gamma-\alpha}{\vartheta}+1}{\tan\dfrac{\pi}{2}\dfrac{\gamma+\alpha}{\vartheta}\Big/\tan\dfrac{\pi}{2}\dfrac{\gamma-\alpha}{\vartheta}-1}}\right]$ $Z_i = \dfrac{\pi\eta}{8}\Big/\ln\left[2\sqrt{\tan\dfrac{\pi}{2}\dfrac{\gamma+\alpha}{\vartheta}\Big/\tan\dfrac{\pi}{2}\dfrac{\gamma-\alpha}{\vartheta}}\right]$ $Z_m = \eta/4$	116 D 16
20		$Z_h = \dfrac{\eta}{2\pi}\ \ln\left[2\sqrt{\dfrac{\tan\dfrac{\gamma+\alpha}{4}\Big/\tan\dfrac{\gamma-\alpha}{4}+1}{\tan\dfrac{\gamma+\alpha}{4}\Big/\tan\dfrac{\gamma-\alpha}{4}-1}}\right]$ $Z_i = \dfrac{\pi\eta}{8}\Big/\ln\left[2\sqrt{\tan\dfrac{\gamma+\alpha}{4}\Big/\tan\dfrac{\gamma-\alpha}{4}}\right]$ $Z_m = \eta/4$	117 D17
21		$Z = \eta/4$	117

Cons. no.	System	Formulas	Equation & Dia. no.
22		$$Z_h = \frac{2\eta}{N\pi}\ln\left[2\sqrt{\tan\frac{N}{2}(\gamma+\alpha)\big/\tan\frac{N}{2}(\gamma-\alpha)}+1\right]$$ $$Z_i = \frac{\pi\eta}{2N}\Big/\ln\left[2\sqrt{\tan\frac{N}{2}(\gamma+\alpha)\big/\tan\frac{N}{2}(\gamma-\alpha)}-1\right]$$ $$Z_m = \eta/N,\ N\geq 2$$	119
23		$$Z_h = \frac{\eta}{N\pi}\ln\left[2\sqrt{\tan\frac{N}{2}(\gamma+\alpha)\big/\tan\frac{N}{2}(\gamma-\alpha)}+1\right]$$ $$Z_i = \frac{\pi\eta}{4N}\Big/\ln\left[2\sqrt{\tan\frac{N}{2}(\gamma+\alpha)\big/\tan\frac{N}{2}(\gamma-\alpha)}-1\right]$$ $$Z_m = \eta/2N,\ N\geq 2$$	119
24		$$Z_h = \frac{\eta}{2\pi}\ln\left[2\frac{r'}{r}\sqrt{\frac{1+\sqrt{1+r^2/a^2}}{1+\sqrt{1+r'^2/a^2}}}+1\right]$$ $$Z_i = \frac{\pi\eta}{8}\Big/\ln\left[2\frac{r'}{r}\sqrt{\frac{1+\sqrt{1+r^2/a^2}}{1+\sqrt{1+r'^2/a^2}}}-1\right]$$ $$Z_m = \eta/4$$	124 / D 18

Cons. no.	System	Formulas	Equation & Dia. no.
25		$$Z_h = \frac{\eta}{2\pi}\ln\left[2\frac{\sqrt{\frac{a}{r}(1+\sqrt{1+r^2/a^2})}+1}{\sqrt{\frac{a}{r}(1+\sqrt{1+r^2/a^2})}-1}\right]$$ $$Z_l = \frac{\pi\eta}{8}/\ln\left[2\sqrt{\frac{a}{r}(1+\sqrt{1+r^2/a^2})}\right]$$ $$Z_m = \eta/4.$$	125
26		$$Z_h = \frac{\eta}{2\pi}\ln\left[2\frac{\sqrt{r'+\sqrt{a^2+r'^2}}}{r+\sqrt{a^2+r^2}}+1\right]$$ $$Z_l = \frac{\pi\eta}{8}/\ln\left[2\sqrt{\frac{r'+\sqrt{a^2+r'^2}}{r+\sqrt{a^2+r^2}}}-1\right]$$ $$Z_m = \eta/4$$	128
27		$$Z_h = \frac{\eta}{2\pi}\ln\left[2\frac{(r'/r)^{\frac{1}{2n}}+1}{(r'/r)^{\frac{1}{2n}}-1}\right] \quad ; n = \frac{\vartheta}{\pi}$$ $$Z_l = \frac{\pi\eta}{8}/\ln\left[2(r'/r)^{\frac{1}{2n}}\right]$$ $$Z_m = \eta/4$$	130

Cons. no.	System	Formulas	Equation & Dia. no.
28		$Z_h = \dfrac{\eta}{2\pi} \ln\left[2\,\dfrac{\sqrt[4]{r'/r}+1}{\sqrt[4]{r'/r}-1}\right]$ $Z_l = \dfrac{\pi\eta}{8} \Big/ \ln\left[2\sqrt[4]{r'/r}\right]$ $Z_m = \eta/4$	.132 D 20
29		$Z_h = \dfrac{2\eta}{\pi N} \ln\left[2\,\dfrac{(r'/r)^{N/2}+1}{(r'/r)^{N/2}-1}\right]$ $Z_l = \dfrac{\pi\eta}{2N} \Big/ \ln\left[2\,(r'/r)^{N/2}\right]$ $Z_m = \eta/N$	.131
30		$Z_h = \dfrac{\eta}{\pi N} \ln\left[2\,\dfrac{(r'/r)^{N/2}+1}{(r'/r)^{N/2}-1}\right]$ $Z_l = \dfrac{\pi\eta}{4N} \Big/ \ln\left[2\,(r'/r)^{N/2}\right]$ $Z_m = \eta/2N$	131

Cons. no.	System	Formulas	Equation & Dia. no.
31		$$Z_h = \frac{\eta}{2\pi} \ln \left[2 \left(\frac{r'-a}{r'+a} \right)^{1/4} \frac{r+a}{r-a} + 1 \right]$$ $$Z_l = \frac{\pi\eta}{8} \Big/ \ln \left[2 \left(\frac{r'-a}{r'+a} \right)^{1/4} \frac{r+a}{r-a} - 1 \right]$$ $$Z_m = \eta^{1/4}$$	140 D 21
32		$$Z_h = \frac{2\eta}{\pi} \ln \left[2 \left(\frac{r'-a}{r'+a} \right)^{1/4} \frac{r+a}{r-a} + 1 \right]$$ $$Z_l = \frac{\pi\eta}{2} \Big/ \ln \left[2 \left(\frac{r'-a}{r'+a} \right)^{1/4} \frac{r+a}{r-a} - 1 \right]$$ $$Z_m = \eta$$	141 D 22
33		$$Z_h = \frac{2}{\pi} \ln \left[2\sqrt{D'/D} \right]$$ $$Z_l = \frac{\pi\eta}{4} \Big/ \ln \left[2 \frac{\sqrt{D'/D}+1}{\sqrt{D'/D}-1} \right]$$ $$Z_m = \eta/2$$	38 D 23

Cons. no.	System	Formulas	Equation & Dia. no.
34		$Z_h = \dfrac{\eta}{2\pi}\ln\left[2\coth\dfrac{\pi u}{2h}\right]$ $Z_l = \dfrac{\pi\eta}{8}\Big/\ln\left[2e^{\frac{\pi u}{h}}\right]$ $Z_m = \eta/4$	176 D 24
35		$Z_h = \dfrac{\eta}{2\pi}\ln\left[2\cot\dfrac{\pi}{2h}\left(\dfrac{h}{2}-v_1\right)\right] = \dfrac{\eta}{2\pi}\ln\left[2\cot\left(\dfrac{\pi}{4}-\dfrac{\pi v_1}{2h}\right)\right]$ $Z_l = \dfrac{\pi\eta}{8}\Big/\ln\left[2\cot\dfrac{\pi v_1}{2h}\right]$ $Z_m = \eta/4$	174 D 25
36		$Z_h = \dfrac{\eta}{2\pi}\ln\left[2\dfrac{\sqrt{\tan\frac{\pi}{2}\frac{v_2}{h}/\tan\frac{\pi}{2}\frac{v_1}{h}}+1}{\sqrt{\tan\frac{\pi}{2}\frac{v_2}{h}/\tan\frac{\pi}{2}\frac{v_1}{h}}-1}\right]$ $Z_l = \dfrac{\pi\eta}{8}\Big/\ln\left[2\sqrt{\tan\frac{\pi}{2}\frac{v_2}{h}/\tan\frac{\pi}{2}\frac{v_1}{h}}\right]$ $Z_m = \eta/4$	173 D 26

Cons. no.	System	Formulas	Equation & Dia. no.
37		$$Z_h = \frac{\eta}{\pi} \ln\left[2\sqrt{\tan\frac{\pi}{2}\frac{v_2}{h}\Big/\tan\frac{\pi}{2}\frac{v_1}{h}} + 1\right]$$ $$Z_l = \frac{\pi\eta}{4}\Big/\ln\left[2\sqrt{\tan\frac{\pi}{2}\frac{v_2}{h}\Big/\tan\frac{\pi}{2}\frac{v_1}{h}} - 1\right]$$ $$Z_m = \eta/2$$	1.173 D 27
38		$$Z_h = \frac{\eta}{\pi} \ln\left[2\sqrt{\sinh\frac{\pi u_1}{h}\Big/\sinh\frac{\pi u_2}{h}} + 1\right]$$ $$Z_l = \frac{\pi\eta}{4}\Big/\ln\left[2\sqrt{\sinh\frac{\pi u_1}{h}\Big/\sinh\frac{\pi u_2}{h}} - 1\right]$$ $$Z_m = \eta/2$$	.179 D 28
39		$$Z_h = \frac{\eta}{2\pi} \ln\left[2\left(\tan\frac{\gamma+\alpha}{2}\Big/\tan\frac{\gamma-\alpha}{2}\right)^{1/4} + 1\right]$$ $$Z_l = \frac{\pi\eta}{8}\Big/\ln\left[2\left(\tan\frac{\gamma+\alpha}{2}\Big/\tan\frac{\gamma-\alpha}{2}\right)^{1/4} - 1\right]$$ $$Z_m = \eta/4$$	.135 D 29

Cons. no.	System	Formulas	Equation & Dia. no.
40		$$Z_h = \frac{\eta}{2\pi}\ \ln\left[2\ \frac{\sqrt{\tan\left(\frac{\pi}{4}+\alpha\right)}+1}{\sqrt{\tan\left(\frac{\pi}{4}+\alpha\right)}-1}\right]$$ $$Z_i = \frac{\pi\eta}{8}\Big/\ln\left[2\sqrt{\tan\left(\frac{\pi}{4}+\alpha\right)}\right]$$ $$Z_m = \eta/4$$	135
41		$$Z_h = \frac{\eta}{\pi}\ \ln\left[2\ \frac{\sqrt[4]{(R'^2-A^2)/(R^2-A^2)}+1}{\sqrt[4]{(R'^2-A^2)/(R^2-A^2)}-1}\right]$$ $$Z_i = \frac{\pi\eta}{4}\Big/\ln\left[2\sqrt[4]{(R'^2-A^2)/(\dot{R}^2-A^2)}\right]$$ $$Z_m = \eta/2$$	182 D 30
42		$$Z_h = \frac{\eta}{4\pi}\ \ln\left[2\ \frac{\left(\frac{R'^2-A^2}{R'^2}\ \frac{R^2}{R^2-A^2}\right)^{1/4}+1}{\left(\frac{R'^2-A^2}{R'^2}\ \frac{R^2}{R^2-A^2}\right)^{1/4}-1}\right]$$ $$Z_i = \frac{\pi\eta}{16}\Big/\ln\left[2\left(\frac{R'^2-A^2}{R'^2}\ \frac{R^2}{R^2-A^2}\right)^{1/4}\right]$$ $$Z_m = \eta/8$$	185 D 31

Cons. no.	System	Formulas	Equation & Dia. no.
43		$Z_h = \dfrac{\eta}{\pi} \ln\left[2\sqrt[4]{(R'^2 - A^2)/(R^2 - A^2)}\right]$	183
		$Z_l = \dfrac{\pi\eta}{4} \Big/ \ln\left[2\,\dfrac{\sqrt[4]{(R'^2 - A^2)/(R^2 - A^2)}+1}{\sqrt[4]{R'^2 - A^2)/(R^2 - A^2)}-1}\right]$	D 33
		$Z_m = \eta/2$	
44		$Z_h = \dfrac{4\eta}{\pi} \ln\left[2\left(\dfrac{R'^2 - A^2}{R'^2}\,\dfrac{R^2}{R^2 - A^2}\right)^{1/4}\right]$	186
		$Z_l = \pi\eta \Big/ \ln\left[2\,\dfrac{\left(\dfrac{R'^2 - A^2}{R'^2}\,\dfrac{R^2}{R^2 - A^2}\right)^{1/4}+1}{\left(\dfrac{R'^2 - A^2}{R'^2}\,\dfrac{R^2}{R^2 - A^2}\right)^{1/4}-1}\right]$	D 32
		$Z_m = 2\eta$	
45		$Z_h = \dfrac{\eta}{2\pi} \ln\left[2\,\dfrac{r'}{r}\right]$	90
		$Z_l = \dfrac{\pi\eta}{8} \Big/ \ln\left[2\,\dfrac{r'+r}{r'-r}\right]$	D 7
		$Z_m = \eta/4$	

Cons. no.	System	Formulas	Equation & Dia. no.
46		$$Z_h = \frac{\eta}{\pi} \ln\left[2\sqrt{\frac{r'}{r}\,\frac{a^2 - r^2}{a^2 - r'^2}} + 1\right]$$ $$Z_1 = \frac{\pi\eta}{4} \Big/ \ln\left[2\sqrt{\frac{r'}{r}\,\frac{a^2 - r^2}{a^2 - r'^2}} - 1\right]$$ $$Z_m = \eta/2$$	120c
47		$$Z_h = \frac{\eta}{2\pi}\,\text{arcosh}\,\frac{\sin\pi\frac{\gamma_2}{\vartheta_2}}{\sin\pi\frac{\beta_2}{\vartheta_2}} \quad ; \quad \vartheta_2 - \beta_2 > \gamma_2 > \beta_2$$	201
48		$$Z_h = \frac{\eta}{2\pi}\,\text{arcosh}\,\frac{\sin\gamma_2/2}{\sin\beta_2/2}$$	202

Cons. no.	System	Formulas	Equation & Dia. no.
49		$$Z_h = \frac{\eta}{2\pi}\,\text{arcosh}\,\frac{\sin 2\gamma_2}{\sin 2\beta_2}$$ $$= \frac{\eta}{2\pi}\,\text{arcosh}\,\frac{2A_2 B_2}{d_2\sqrt{A_2^2 - B_2^2}}$$	203 .218
50		$$Z_h = \frac{\eta}{2\pi}\,\text{arcosh}\,\frac{\sin\left[\frac{\pi}{4}+\frac{1}{2}\arctan\frac{a(r+r')}{a^2-rr'}\right]}{\sin\left[\frac{1}{2}\arctan\frac{a(r'-r)}{a^2+rr'}\right]}\cdot$$ $$= \frac{\eta}{2\pi}\,\text{arcosh}\,\frac{\sin\left[\frac{\pi}{4}+\frac{1}{2}\arctan\frac{8RD}{4R^2-4D^2+d^2}\right]}{\sin\left[\frac{1}{2}\arctan\frac{4Rd}{4R^2+4D^2-d^2}\right]}$$	205 .207
51		$$Z_h = \frac{\eta}{2\pi}\,\text{arcosh}\,\frac{\sin\left[\frac{1}{2}\arctan\frac{a(r+r')}{a^2-rr'}\right]}{\sin\left[\frac{1}{2}\arctan\frac{a(r'-r)}{a^2+rr'}\right]}$$ $$= \frac{\eta}{2\pi}\,\text{arcosh}\,\frac{\sin\left[\frac{1}{2}\arctan\frac{4LD}{L^2-4D^2+d^2}\right]}{\sin\left[\frac{1}{2}\arctan\frac{2Ld}{L^2+4D^2-d^2}\right]}$$	209 .211

Cons. no.	System	Formulas	Equation & Dia. no.
52		$$Z_h = \frac{\eta}{2\pi}\,\text{arcosh}\,\frac{\sin\left[\dfrac{\pi}{2} - \dfrac{1}{2}\arctan\dfrac{4LD}{L^2 - 4D^2 + d^2}\right]}{\sin\left[\dfrac{1}{2}\arctan\dfrac{2Ld}{L^2 + 4D^2 - d^2}\right]}$$	213
53		$$Z_h = \frac{\eta}{\pi}\,\text{arcosh}\,\frac{A_2 L}{d_2\sqrt{A_2^2 + L^2/4}}$$	219 D34
54		$$Z_h = \frac{\eta}{2\pi}\,\text{arcosh}\,\frac{SL}{d\sqrt{S^2 + L^2}} = \frac{\eta}{2\pi}\,\text{arcosh}\,\frac{SL}{dD}$$	220 D35

Tabelle I

Cons. no.	System	Formulas	Equation & Dia. no.
55		$Z_h = \dfrac{\eta}{N\pi}$ arcosh $1/\sin(N \arcsin d/D)$ Anzahl der Leiter: $2N$	Gl. 225
56		$Z_h = \dfrac{\eta}{\pi}$ arcosh $\dfrac{\sin 2 \arctan \dfrac{a(r'+r)}{a^2 - rr'}}{\sin 2 \arctan \dfrac{a(r'-r)}{a^2 + rr'}}$	Gl. 227 D 36
57		$Z_h = \dfrac{\eta}{\pi}$ arcosh $\dfrac{\sin 2 \arctan \dfrac{a(r'+r)}{a^2 - rr'}}{\sin 2 \arctan \dfrac{a(r'-r)}{a^2 + rr'}}$	Gl. 227 D 36

Tabelle I

Cons. no.	System	Formulas	Equation & Dia. no.
58		$$Z_h = -\sqrt{\frac{\mu}{\epsilon}}\;\frac{1}{2\pi}\ln\frac{dL}{2SD} = \frac{\eta}{2\pi}\ln\frac{2SD}{dL}$$	Gl. 234 D 39
59		$$Z_h = \frac{\eta}{4\pi}\ln\frac{4A\sqrt{L^2+4A^2}}{dL}$$	Gl. 236 D 37
60		$$Z_h = \frac{\eta}{2\pi}\ln\frac{4A\sqrt{A^2+B^2}}{dB}$$	Gl. 237

Cons. no.	System	Formulas	Equation & Dia. no.
51		$$Z_h = \frac{\eta}{\pi} \ln \frac{S\sqrt{S^2+4B^2}}{dB}$$	238
52		$$\frac{h}{b} = \pi/2\left[\frac{\pi\eta}{2Z} - \ln 2 - \ln\left(\frac{\pi\eta}{2Z} - \ln 2\right) - 1\right], \text{ für } 0 \le \frac{Z}{\eta} \le 0.35$$ $$\frac{h}{b} = \frac{1}{8}\, e^{\frac{Z 2\pi}{\eta}}, \text{ for } 0.35 \le \frac{Z}{\eta} \le \infty$$	240 242 243 D55
53		$$\frac{d}{b} = \pi/\left[\frac{\pi\eta}{Z} - \ln 2 - \ln\left(\frac{\pi\eta}{Z} - \ln 2\right) - 1\right], \text{ für } 0 \le \frac{Z}{\eta} \le 0.7$$ $$\frac{d}{b} = \frac{1}{4}\, e^{\frac{Z\pi}{\eta}}, \text{ for } 0.7 \le \frac{Z}{\eta} \le \infty$$	.244 D56

Cons. no.	System	Formulas	Equation & Dia. no.
64		$Z_h = \dfrac{2\eta}{\pi}\,\ln\left[2\,e^{\frac{\pi u}{h}}\right]$ $Z_l = \dfrac{\pi\eta}{2}\Big/\ln\left[2\coth\dfrac{\pi u}{2h}\right]$ $Z_m = \eta$	247 D 38
65		$\sqrt{\varepsilon_{eff}} = \dfrac{\sqrt{\varepsilon_{r1}}+\sqrt{\varepsilon_{r2}}}{2}$	.255

TABLE II

CONICAL TRANSMISSION LINES

Comments: The relative error of the approximations in which the parameters Z_h, Z_1, and Z_m are indicated, is always smaller than $2.4 \cdot 10^{-3}$. See Equation (77). The boundary between the region for high impedance values Z_h and the region for low impedance values Z_1 is given by Z_m. The constant η is defined in Equation (15).

Cons. no.	System	Formulas	Equation & Dia. no.
1		$Z = \dfrac{\eta}{2\pi} \ln \dfrac{\tan \vartheta_1/2}{\tan \vartheta_2/2}$	35a D 40
2		$Z = \dfrac{\eta}{2\pi} \ln \cot \vartheta/2$	35 b D 41
3		$Z = \dfrac{\eta}{2\pi} \operatorname{arcosh} \dfrac{\cos \vartheta_1 \cos \vartheta_2 - \cos \alpha}{\sin \vartheta_1 \sin \vartheta_2}$	.71 D 42 D 43 D 44

Cons. no.	System	Formulas	Equation & Dia. no.
4		$$Z_h = \frac{\eta}{2\pi}\ln\left[2\,\frac{\tan\alpha_1/2}{\tan\alpha_2/2}\right]$$ $$Z_l = \frac{\pi\eta}{8}\Big/\ln\left[2\,\frac{\tan\alpha_1/2+\tan\alpha_2/2}{\tan\alpha_1/2-\tan\alpha_2/2}\right]$$ $$Z_m = \eta/4$$	73 34 D 45
5		$$Z_h = \frac{\eta}{2\pi}\ln\left[2\cot\alpha/2\right]; \quad \frac{\pi}{4}\geq\alpha\geq 0$$ $$Z_l = \frac{\pi\eta}{8}\Big/\ln\left[2\cot\left(\frac{\pi}{4}-\frac{\alpha}{2}\right)\right]; \quad \frac{\pi}{2}\geq\alpha\geq\frac{\pi}{4}$$ $$Z_m = \eta/4; \quad \alpha_m = \pi/4$$	74 85 D 11
6		$$Z_h = \frac{\eta}{\pi}\ln\left[2\cot\alpha/2\right]; \quad \infty\geq Z_h\geq Z_m; \quad \frac{\pi}{4}\geq\alpha\geq 0; \quad \frac{\Delta Z}{Z}\leq 0,237\%$$ $$Z_l = \frac{\pi\eta}{4}\Big/\ln\left[2\cot\left(\frac{\pi}{4}-\frac{\alpha}{2}\right)\right]; \quad Z_m\geq Z_l\geq 0; \quad \frac{\pi}{2}\geq\alpha\geq\frac{\pi}{4}; \quad \frac{\Delta Z}{Z}\leq 0,236\%$$ $$Z_m = \eta/2; \quad \alpha_m = \pi/4$$	75 76 79 D 9

Cons. no.	System	Formulas	Equation & Dia. no.
7		$$Z_h = \frac{\eta}{2\pi}\ln\left[2\tan\alpha_1/2\right]$$ $$Z_l = \frac{\pi\eta}{8}\Big/\ln\left[2\cot\left(\frac{\alpha_1}{2}-\frac{\pi}{4}\right)\right]$$ $$Z_m = \eta/4$$	95 D46
8		$$Z_h = \frac{\eta}{2\pi}\ln\left[2\sqrt{\tan\frac{\gamma+\alpha}{2}\Big/\tan\frac{\gamma-\alpha}{2}}+1\right]$$ $$Z_l = \frac{\pi\eta}{8}\Big/\ln\left[2\sqrt{\tan\frac{\gamma+\alpha}{2}\Big/\tan\frac{\gamma-\alpha}{2}}-1\right]$$ $$Z_m = \eta/4$$	100 D14
9		$$Z_h = \frac{\eta}{\pi}\ln\left[2\sqrt{\tan\frac{\gamma+\alpha}{2}\Big/\tan\frac{\gamma-\alpha}{2}}+1\right]$$ $$Z_l = \frac{\pi\eta}{4}\Big/\ln\left[2\sqrt{\tan\frac{\gamma+\alpha}{2}\Big/\tan\frac{\gamma-\alpha}{2}}-1\right]$$ $$Z_m = \eta/2$$	101 D15

Cons. no.	System	Formulas	Equation & Dia. no.
10		$$Z_h = \frac{\eta}{2\pi}\ln\left[2\sqrt{\frac{\tan\frac{\gamma+\alpha}{2}/\tan\frac{\gamma-\alpha}{2}+1}{\tan\frac{\gamma+\alpha}{2}/\tan\frac{\gamma-\alpha}{2}-1}}\right]$$ $$Z_l = \frac{\pi\eta}{8}/\ln\left[2\sqrt{\tan\frac{\gamma+\alpha}{2}/\tan\frac{\gamma-\alpha}{2}}\right]$$ $$Z_m = \eta/4$$	109 D 47
11		$$Z_h = \frac{\eta}{2\pi}\ln\left[2\sqrt{\frac{\left(\tan\frac{\gamma+\beta}{2}-\tan\frac{\alpha}{2}\right)\left(\tan\frac{\gamma-\beta}{2}+\tan\frac{\alpha}{2}\right)}{\left(\tan\frac{\gamma+\beta}{2}+\tan\frac{\alpha}{2}\right)\left(\tan\frac{\gamma-\beta}{2}-\tan\frac{\alpha}{2}\right)}+1}\middle/\sqrt{\frac{\left(\tan\frac{\gamma+\beta}{2}-\tan\frac{\alpha}{2}\right)\left(\tan\frac{\gamma-\beta}{2}+\tan\frac{\alpha}{2}\right)}{\left(\tan\frac{\gamma+\beta}{2}+\tan\frac{\alpha}{2}\right)\left(\tan\frac{\gamma-\beta}{2}-\tan\frac{\alpha}{2}\right)}-1}\right]$$ $$Z_l = \frac{\pi\eta}{8}/\ln\left[2\sqrt{\frac{\left(\tan\frac{\gamma+\beta}{2}-\tan\frac{\alpha}{2}\right)\left(\tan\frac{\gamma-\beta}{2}+\tan\frac{\alpha}{2}\right)}{\left(\tan\frac{\gamma+\beta}{2}+\tan\frac{\alpha}{2}\right)\left(\tan\frac{\gamma-\beta}{2}-\tan\frac{\alpha}{2}\right)}}\right]$$ $$Z_m = \eta/4$$	111 D 48 D 49 D 50
12		$$Z_h = \frac{2\eta}{N\pi}\ln\left[2\sqrt{\frac{\tan\frac{N}{2}(\gamma+\alpha)/\tan\frac{N}{2}(\gamma-\alpha)+1}{\tan\frac{N}{2}(\gamma+\alpha)/\tan\frac{N}{2}(\gamma-\alpha)-1}}\right]$$ $$Z_l = \frac{\pi\eta}{2N}/\ln\left[2\sqrt{\tan\frac{N}{2}(\gamma+\alpha)/\tan\frac{N}{2}(\gamma-\alpha)}\right]$$ $$Z_m = \eta/N, \quad N \geq 2$$	119

Cons. no.	System	Formulas	Equation & Dia. no.
13		$$Z_h = \frac{\eta}{N\pi}\ln\left[2\sqrt{\tan\frac{N}{2}(\gamma+\alpha)/\tan\frac{N}{2}(\gamma-\alpha)} + 1\right]$$ $$Z_l = \frac{\pi\eta}{4N}/\ln\left[2\sqrt{\tan\frac{N}{2}(\gamma+\alpha)/\tan\frac{N}{2}(\gamma-\alpha)} - 1\right]$$ $$Z_m = \eta/2N, \quad N \geq 2$$ $$2N$$	119
14		$$Z_h = \frac{\eta}{2\pi}\ln\left[2\sqrt{\tan\frac{\gamma+\alpha}{4}/\tan\frac{\gamma-\alpha}{4}} + 1\right]$$ $$Z_l = \frac{\pi\eta}{8}/\ln\left[2\sqrt{\tan\frac{\gamma+\alpha}{4}/\tan\frac{\gamma-\alpha}{4}} - 1\right]$$ $$Z_m = \eta/4$$	117 D51 D52
15		$$Z_h = \frac{\eta}{2\pi}\ln\left[2\left(\tan\frac{\gamma+\alpha}{2}/\tan\frac{\gamma-\alpha}{2}\right)^{\frac{\pi}{2\delta}} + 1\right]$$ $$Z_l = \frac{\pi\eta}{8}/\ln\left[2\left(\tan\frac{\gamma+\alpha}{2}/\tan\frac{\gamma-\alpha}{2}\right)^{\frac{\pi}{2\delta}} - 1\right]$$ $$Z_m = \eta/4$$	134 D51 D52

Cons. ro.	System	Formulas	Equation & Dia. no.
16		$$Z_h = \frac{\eta}{2\pi}\ln\left[2\,\frac{\left(\tan\frac{\gamma+\alpha}{2}\big/\tan\frac{\gamma-\alpha}{2}\right)^{1/4}+1}{\left(\tan\frac{\gamma+\alpha}{2}\big/\tan\frac{\gamma-\alpha}{2}\right)^{1/4}-1}\right]$$ $$Z_l = \frac{\pi\eta}{8}\Big/\ln\left[2\left(\tan\frac{\gamma+\alpha}{2}\big/\tan\frac{\gamma-\alpha}{2}\right)^{1/4}\right]$$ $$Z_m = \eta/4$$	135 D53 D29
17		$$Z_h = \frac{\eta}{2\pi}\ln\left[2\,\frac{\sqrt{\tan\left(\frac{\pi}{4}+\frac{\alpha}{2}\right)}+1}{\sqrt{\tan\left(\frac{\pi}{4}+\frac{\alpha}{2}\right)}-1}\right]$$ $$Z_l = \frac{\pi\eta}{8}\Big/\ln\left[2\sqrt{\tan\left(\frac{\pi}{4}+\frac{\alpha}{2}\right)}\right]$$ $$Z_m = \eta/4$$	137 D53 D29
18		$$Z_h = \frac{\eta}{\pi}\ln\left[2\sqrt{\cot\frac{\alpha_1}{2}\cot\frac{\alpha_2}{2}}\right]$$ $$Z_l = \frac{\pi\eta}{4}\Big/\ln\left[2\,\frac{1+\sqrt{\tan\frac{\alpha_1}{2}\tan\frac{\alpha_2}{2}}}{1-\sqrt{\tan\frac{\alpha_1}{2}\tan\frac{\alpha_2}{2}}}\right]$$ $$Z_m = \eta/2$$	103 D54

Cons. no.	System	Formulas	Equation & Dia. no.
19		$Z = \eta/2$	248
20		$Z_h = \dfrac{\eta}{2\pi}\ \text{arcosh}\ \dfrac{\sin \gamma/2}{\sin \beta/2}$	202
21		$Z_h = \dfrac{\eta}{2\pi}\ \text{arcosh}\ \dfrac{\sin \pi \dfrac{\gamma}{\vartheta}}{\sin \pi \dfrac{\beta}{\vartheta}}\ ;\quad \vartheta - \beta > \gamma > \beta$	201

Part C

Explanations

The following diagrams were calculated for air-filled field spaces with

$$\eta = \sqrt{\frac{\mu_o}{\epsilon_o}} = 120 \, \pi \text{ Ohm.}$$

The conversion relationships for Z, L', and C' are given in (2). According to this reference, the ordinates of the diagram can be indicated at the same time for all three parameters to scale. For a given Z, the following equations are obtained:

$$C' = \frac{\sqrt{\mu \epsilon}}{Z}$$

$$L' = \sqrt{\mu \epsilon} \cdot Z \quad .$$

If the velocity of light

$$c = \frac{1}{\sqrt{\epsilon_o \cdot \mu_o}} = 299 \, 800 \, \frac{\text{km}}{\text{s}}$$

is approximated by 3.10^8 m/s, then finally obtained for C' and L' are:

$$C' = \frac{10}{3} \cdot \frac{1}{Z} \quad \left[\frac{\text{nF}}{\text{m}}\right] = \frac{10\,000/3}{Z} \quad \left[\frac{\text{pF}}{\text{m}}\right]$$

$$L' = \frac{10}{3} \cdot Z \quad \left[\frac{\text{nH}}{\text{m}}\right] = \frac{Z}{300} \quad \left[\frac{\mu\text{H}}{\text{m}}\right]$$

These are the scales used in the diagrams.

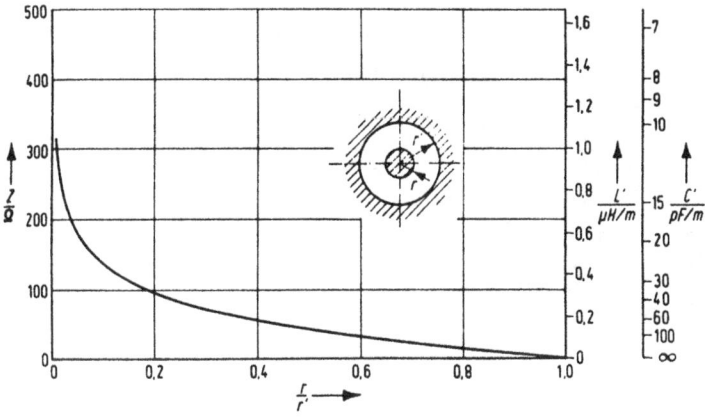

D 1

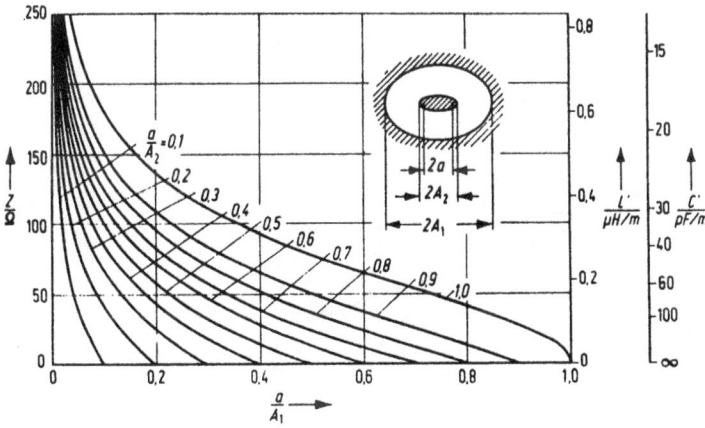

D 2

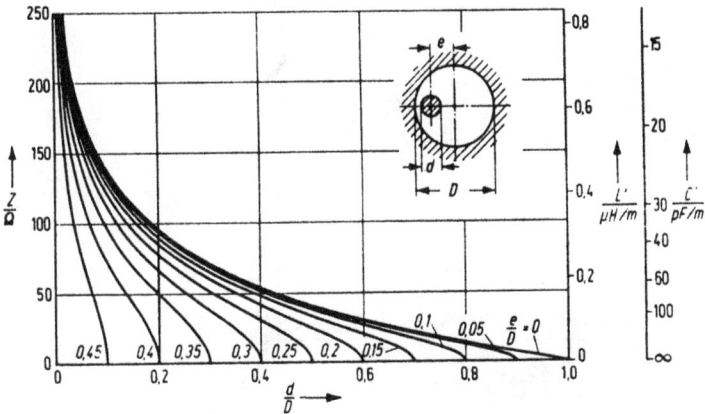

D 3

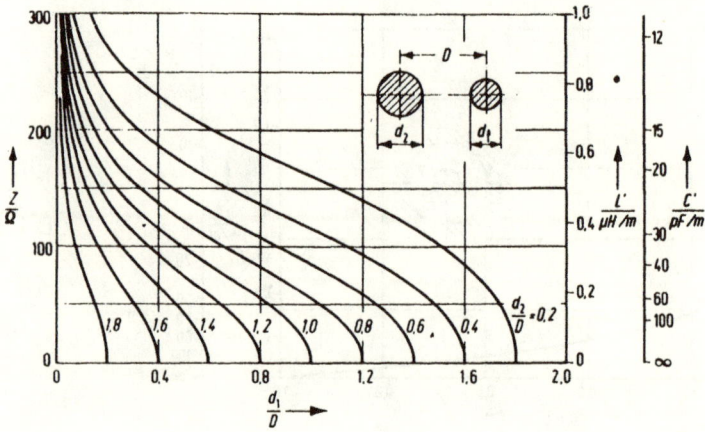

D 4

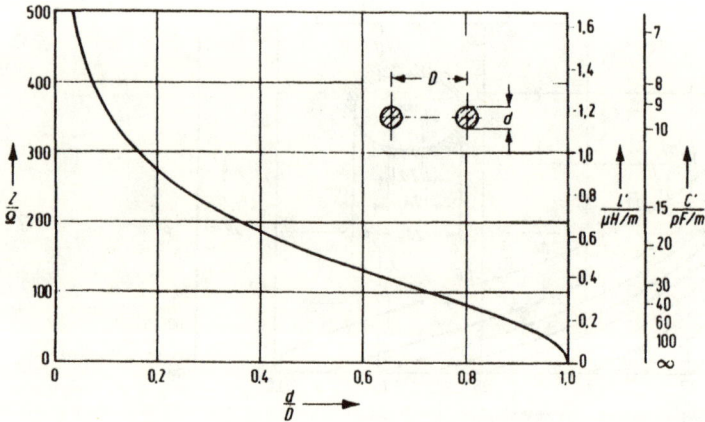

D 5

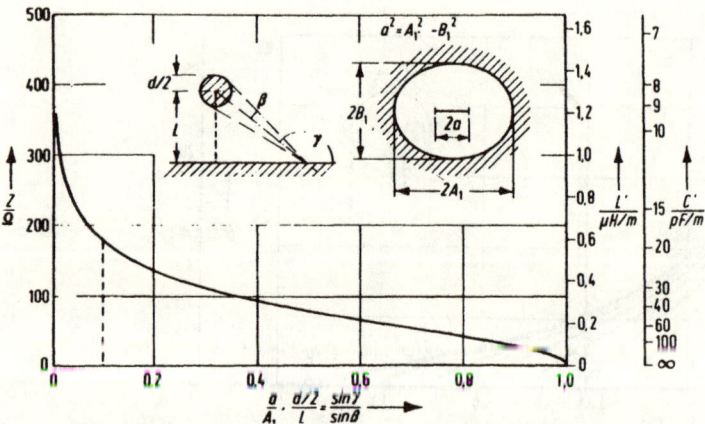

D 6

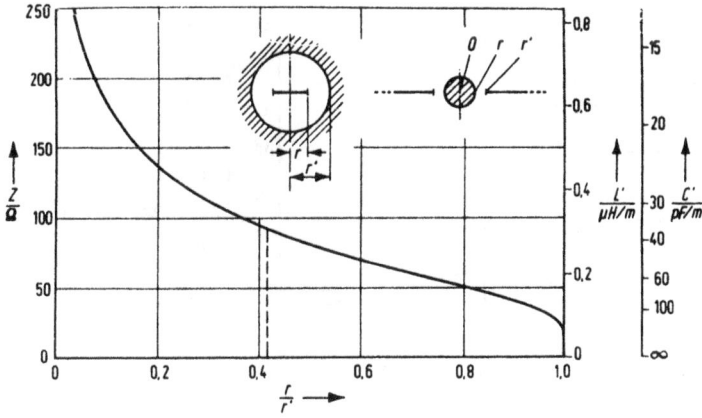

D 7

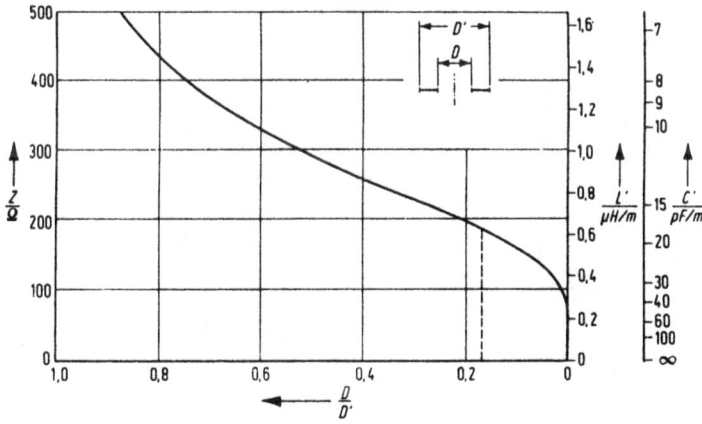

D 8

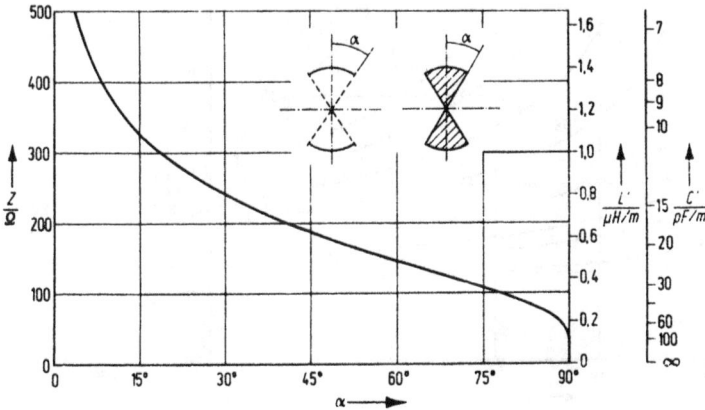

D 9

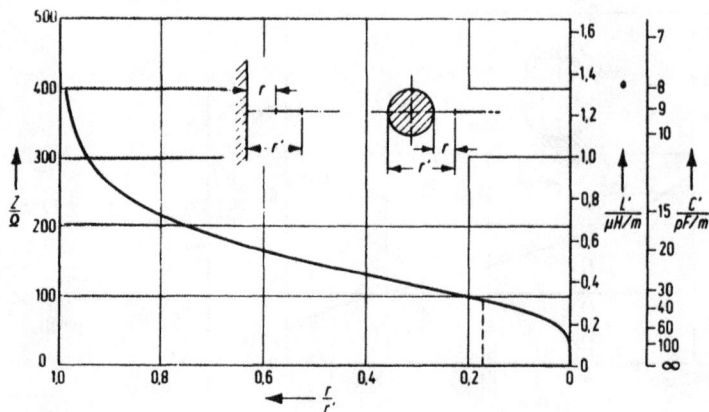

D 10

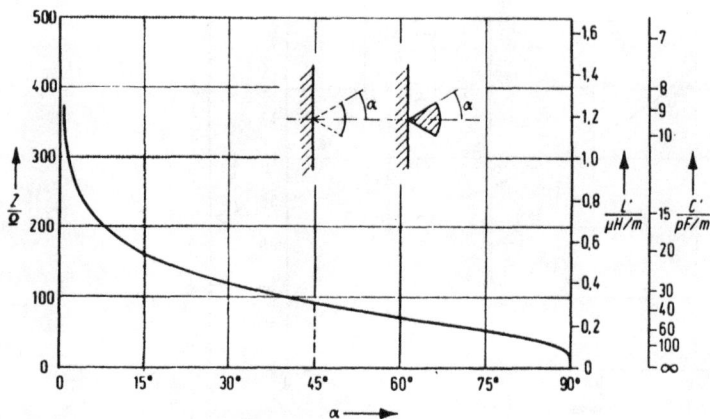

D 11

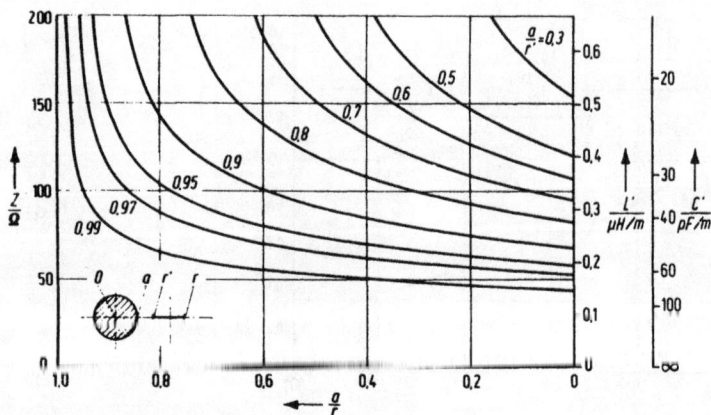

D 12

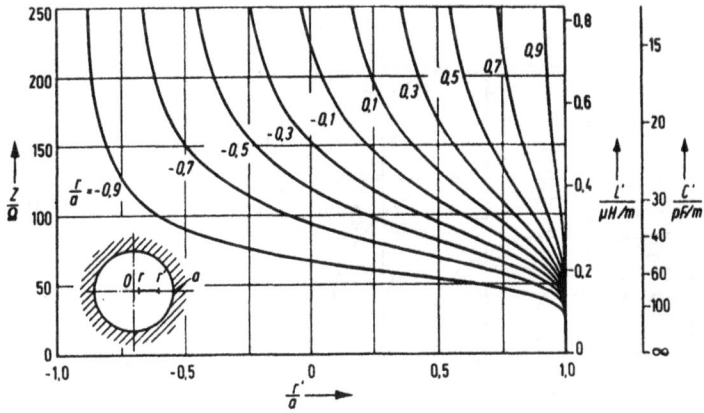

D 13

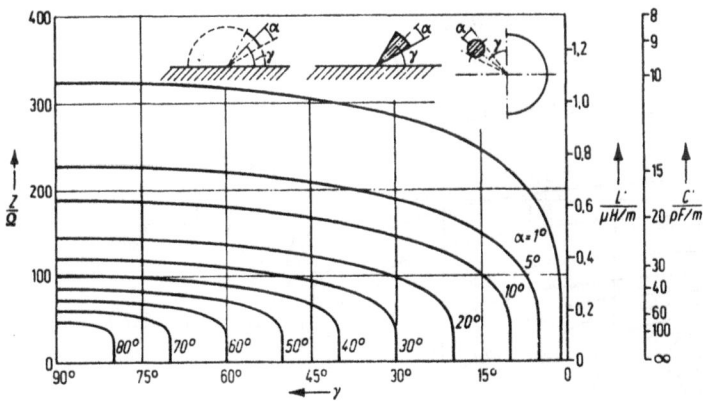

D 14

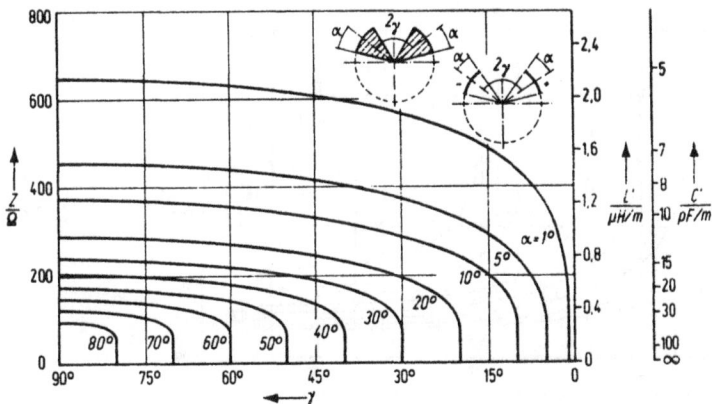

D 15

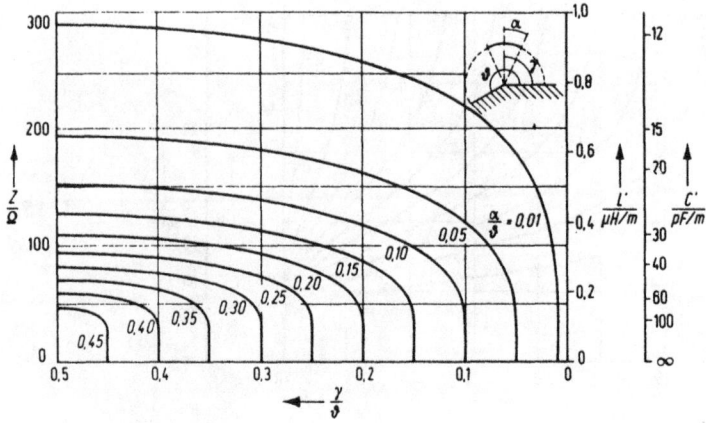

D 16

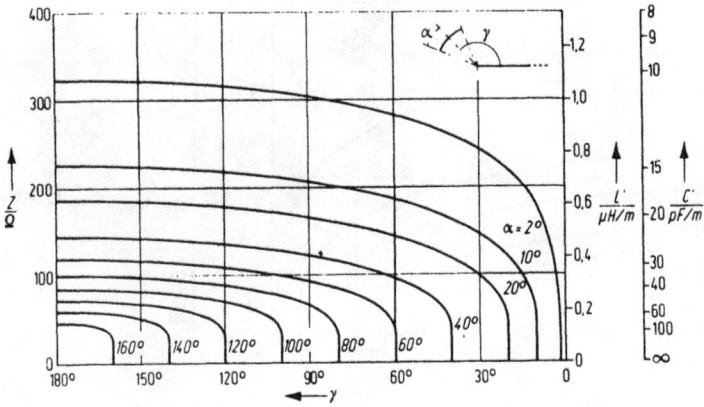

D 17

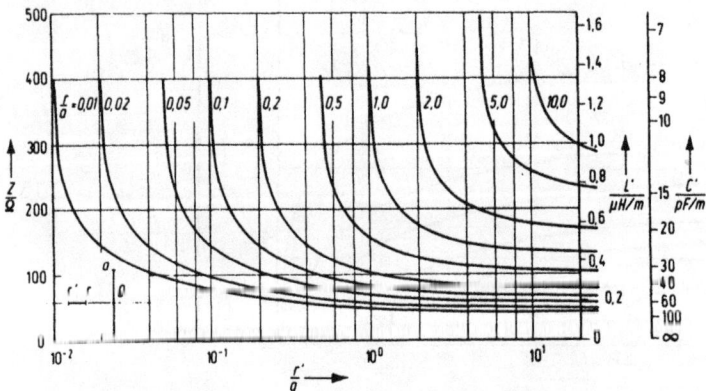

D 18

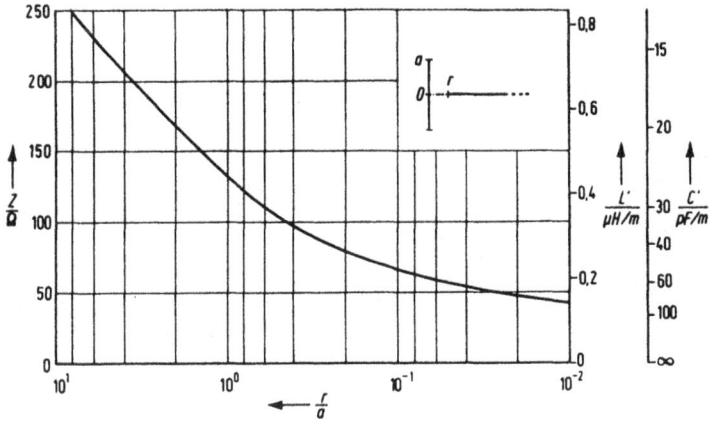

D 19

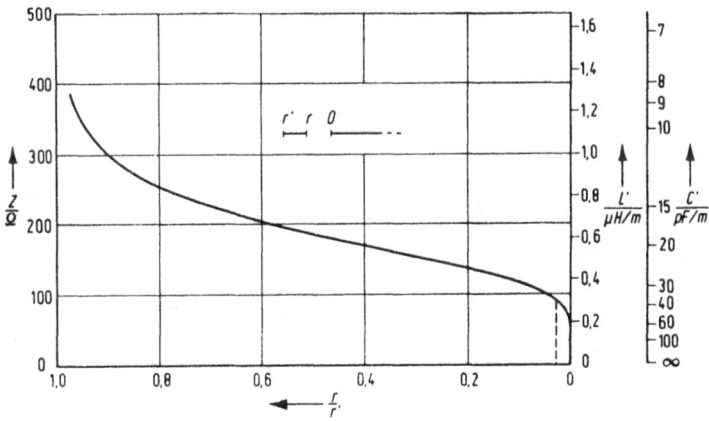

D 20

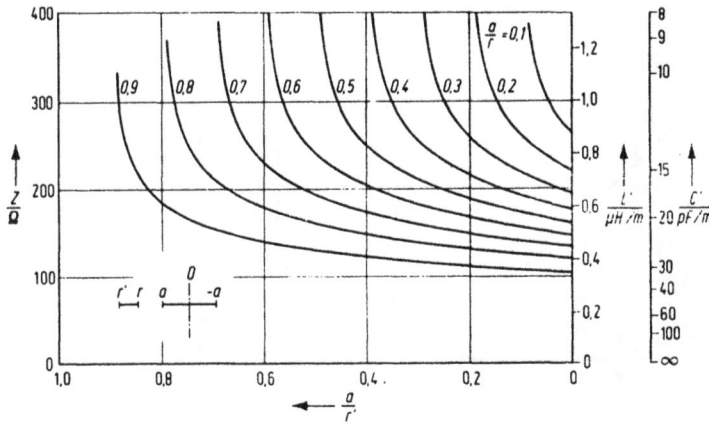

D 21

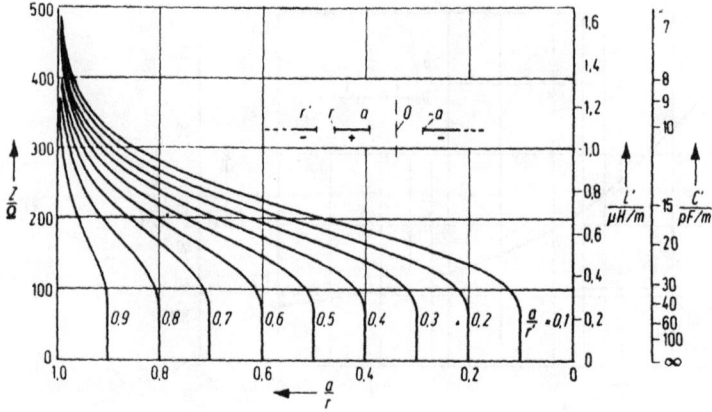

D 22

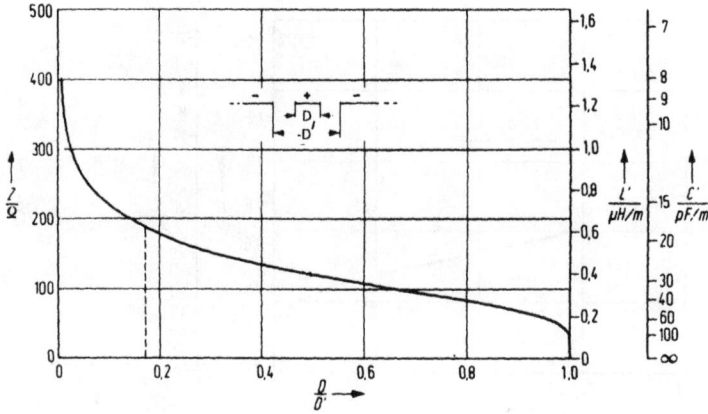

D 23

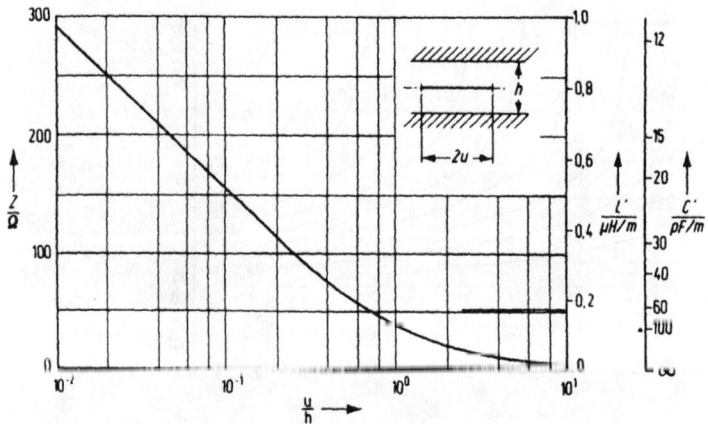

D 24

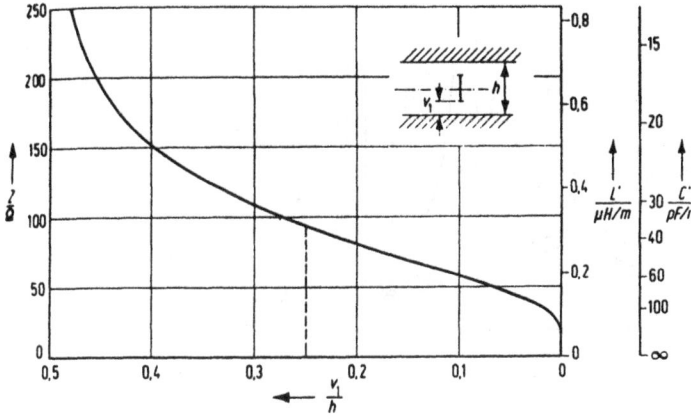

D 25

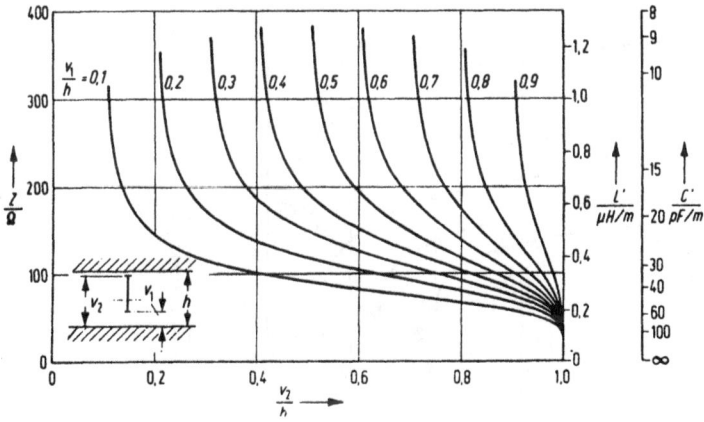

D 26

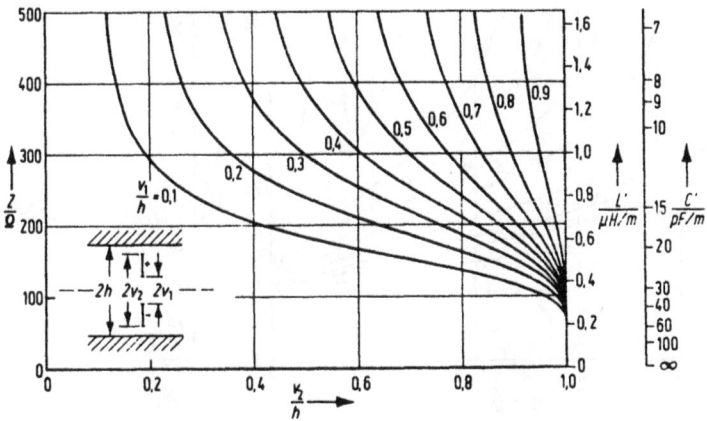

D 27

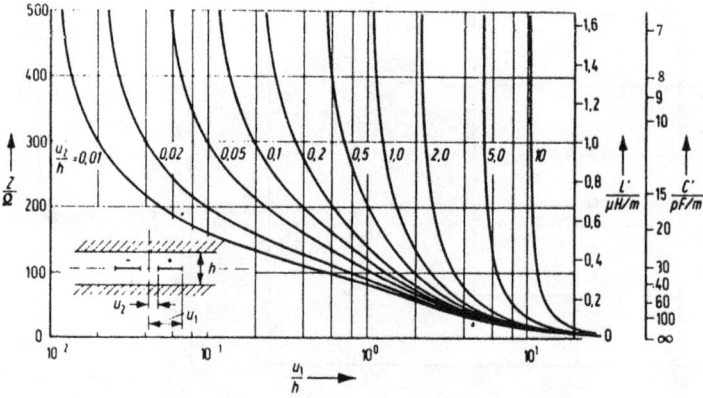

D 28

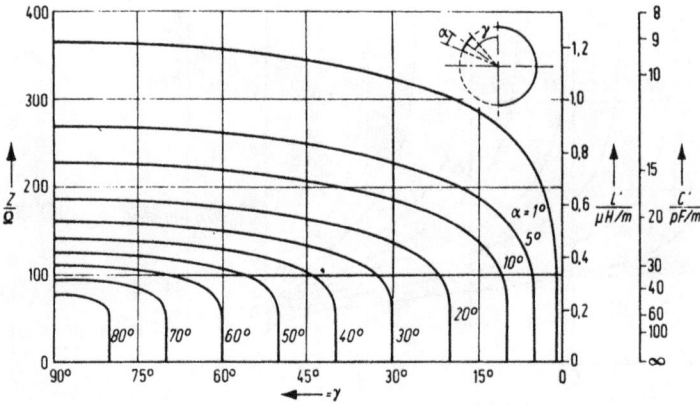

D 29

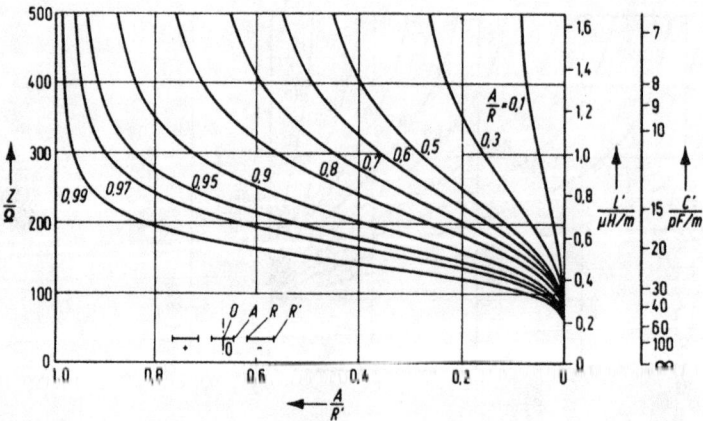

D 30

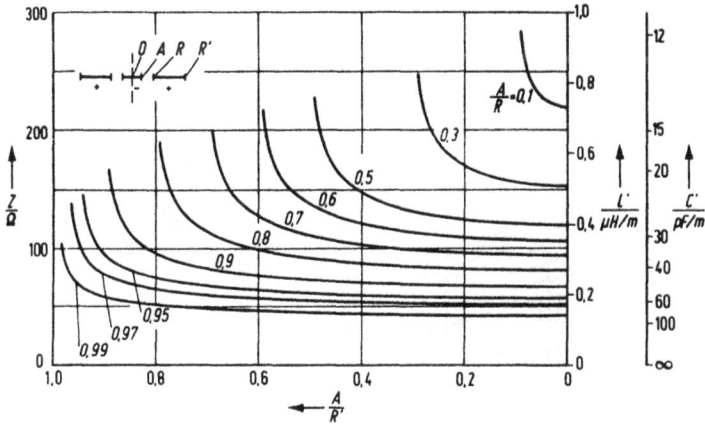

D 31

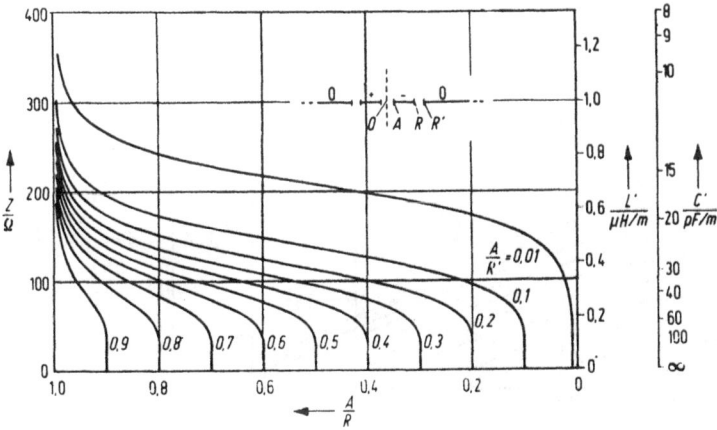

D 32

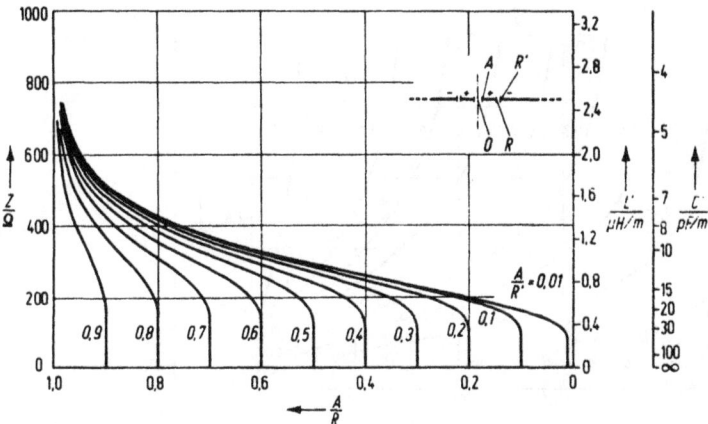

D 33

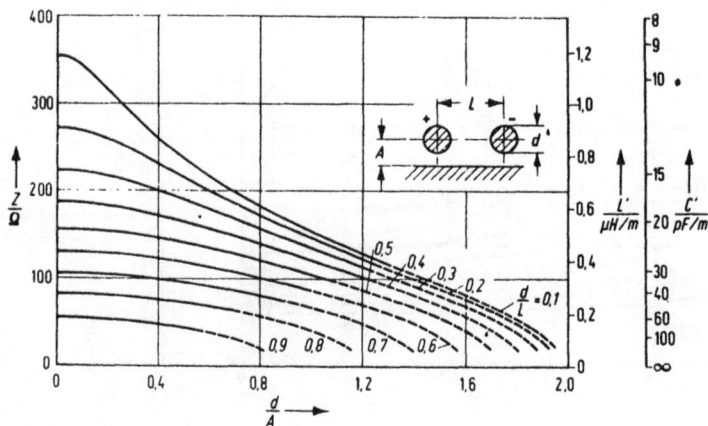

D 34

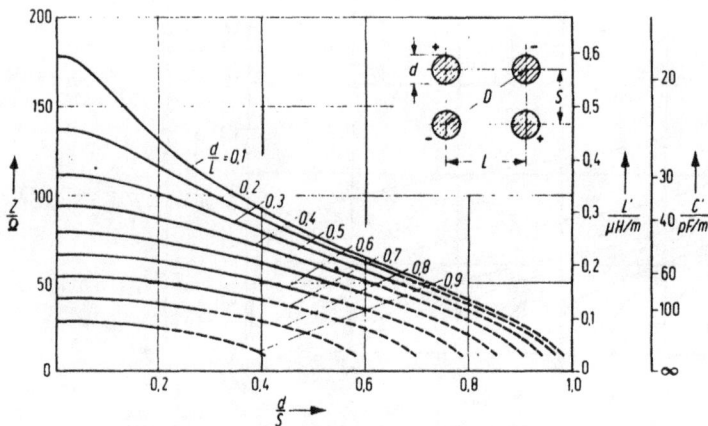

D 35

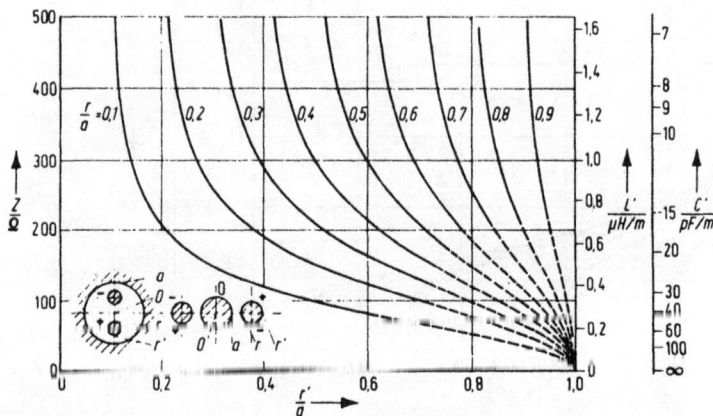

D 36

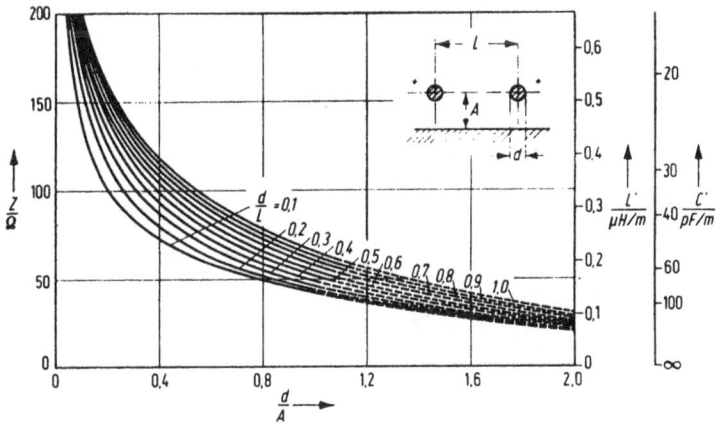

D 37

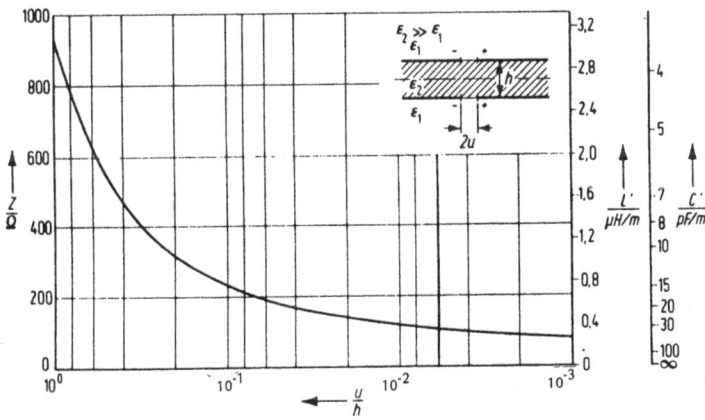

D 38

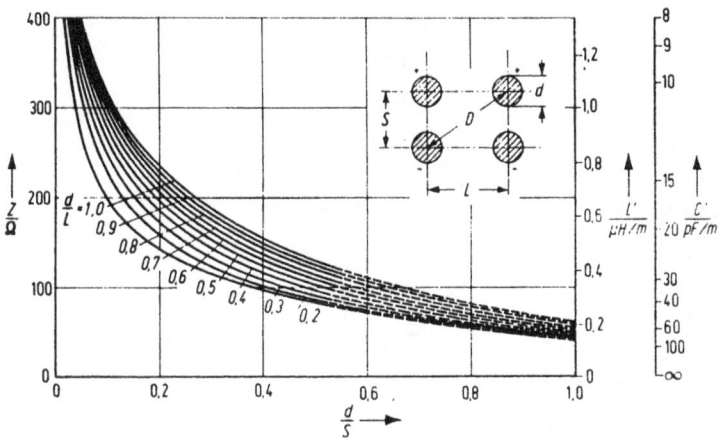

D 39

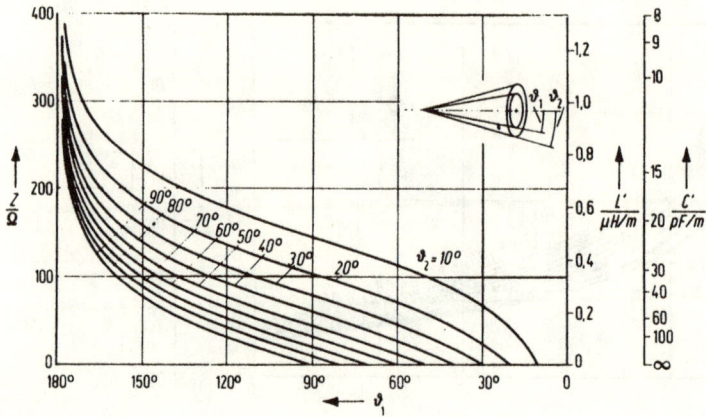

D 40

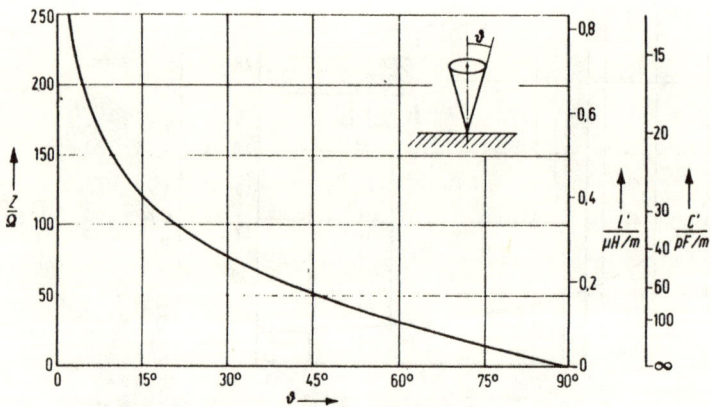

D 41

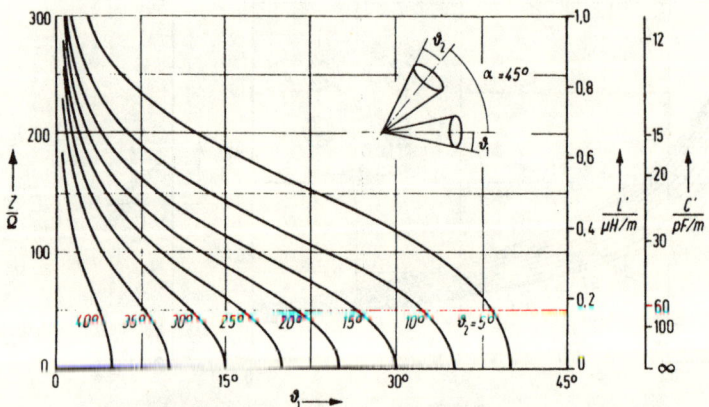

D 42

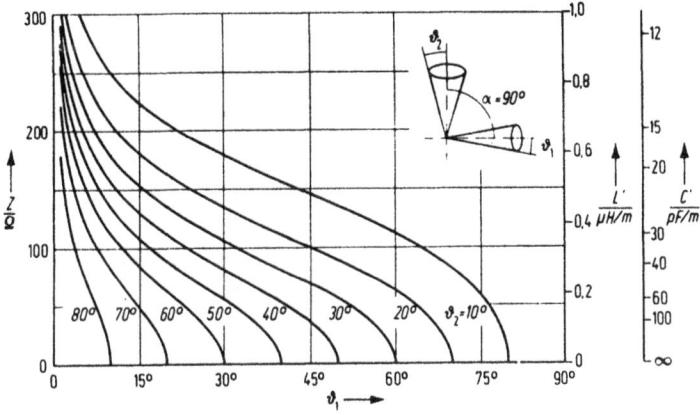

D 43

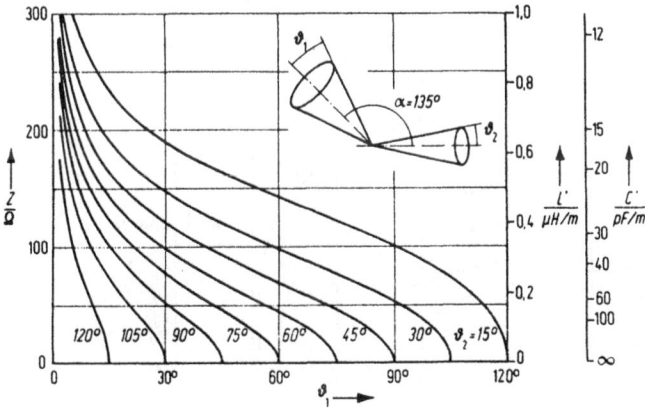

D 44

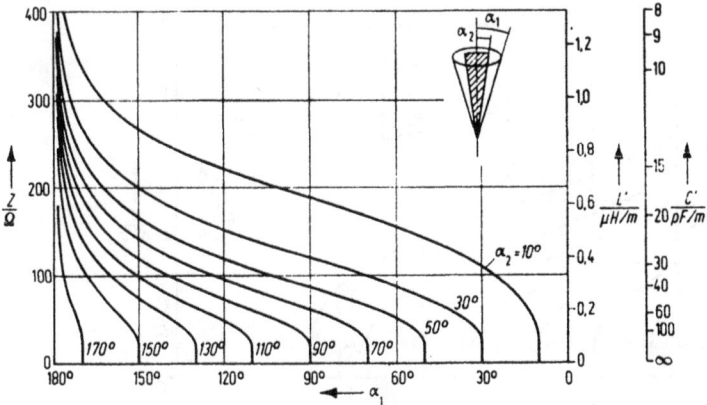

D 45

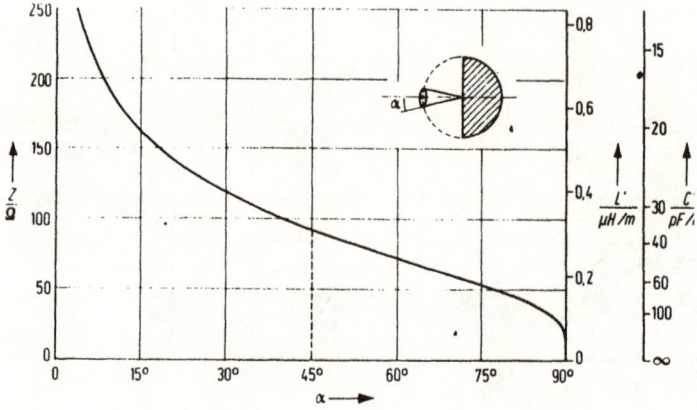

D 46

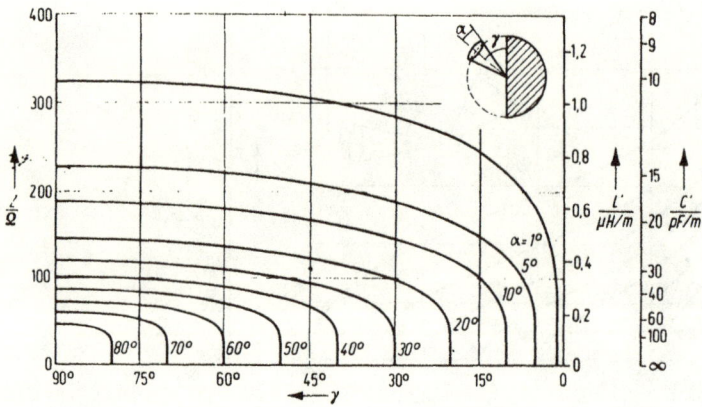

D 47

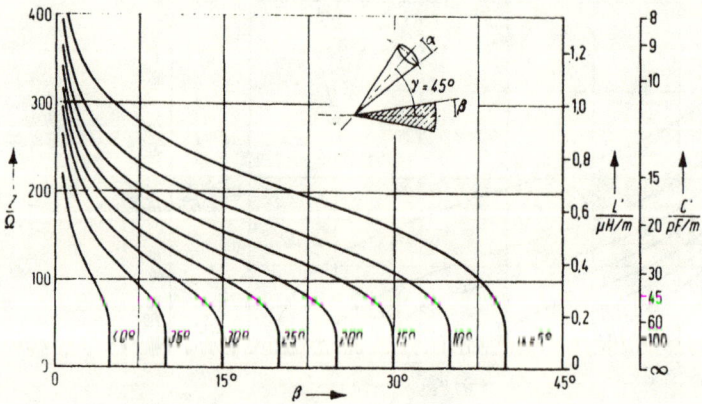

D 48

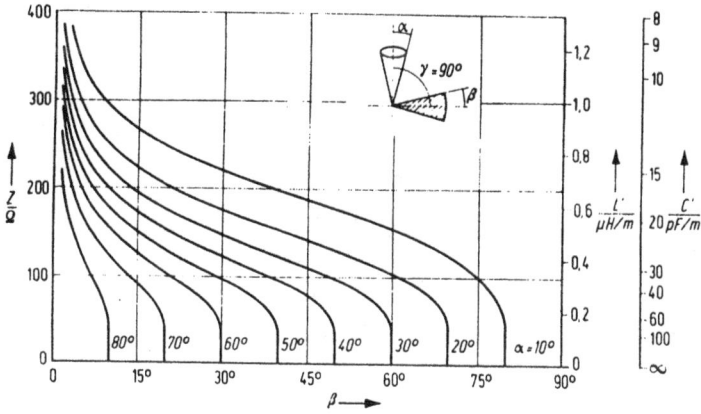

D 49

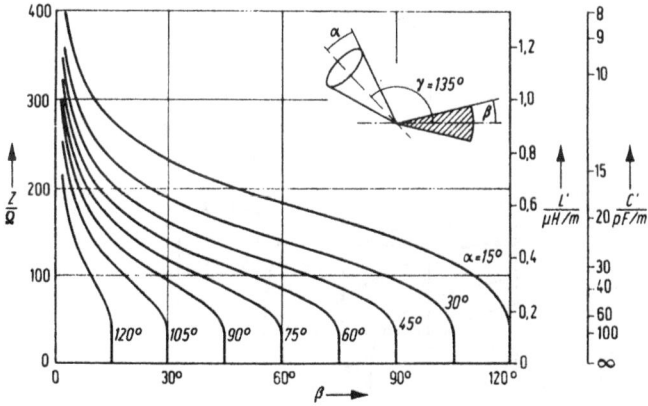

D 50

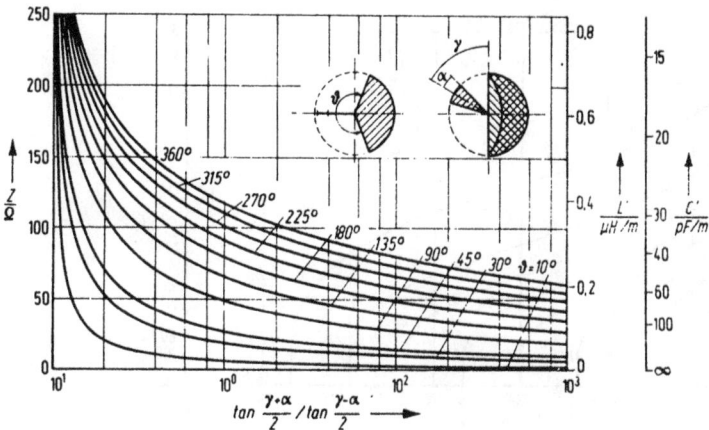

D 51

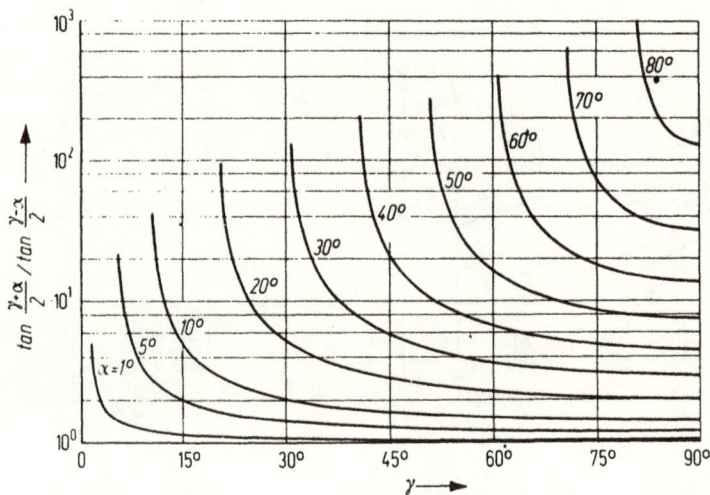

D 52

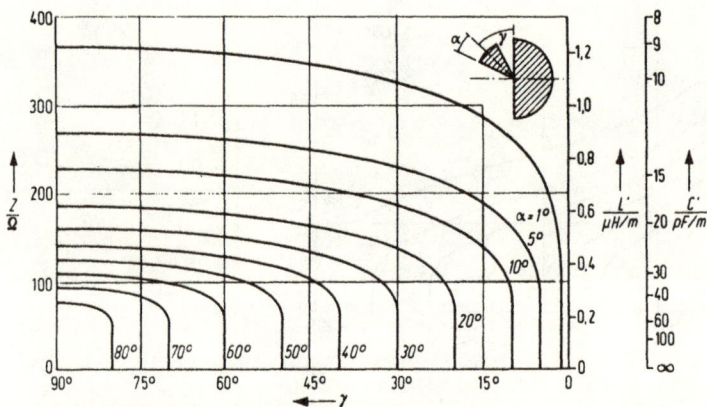

D 53

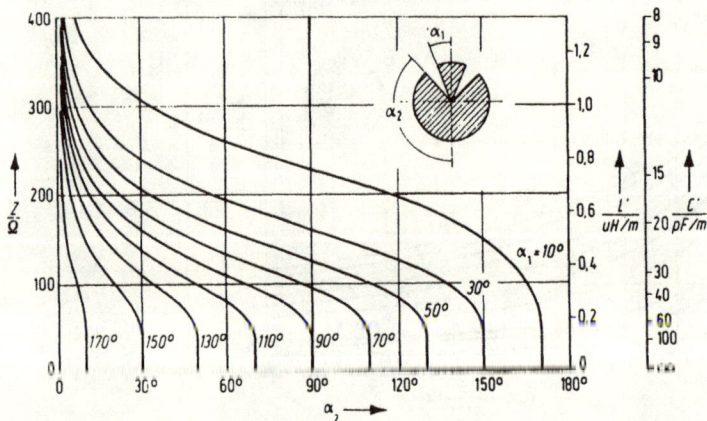

D 54

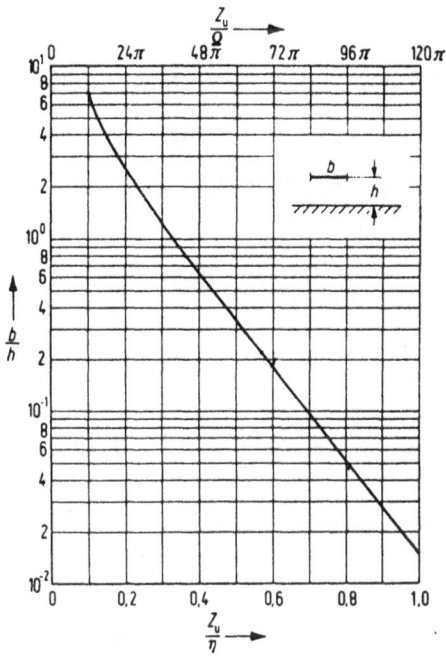

D 55

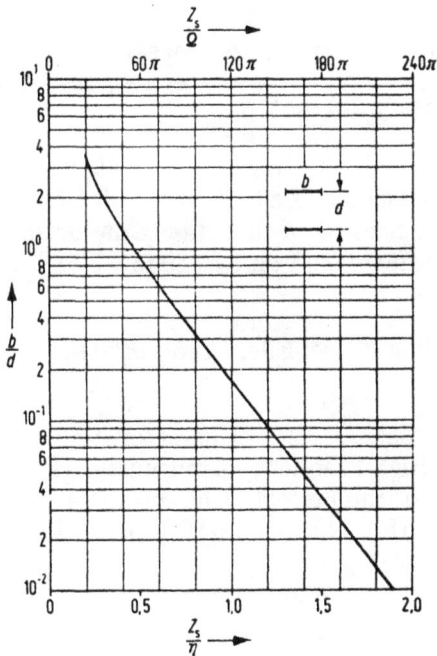

D 56

Part D

References on Transmission Line Calculations

Papers in Chronological Sequence

[1]   Meinke-Gundlach, *Handbook of High Frequency Engineering*, Springer-Verlag.

[2]   Küpfmüller, K., *Introduction to Theoretical Electrical Engineering*, Springer-Verlag.

[3]   Collin, R.E., *Field Theory of Guided Waves*, McGraw-Hill, New York, 1960.

[4]   Megla, G., *Decimeter Wave Technology*, Berlin Union, Stuttgart, 5th Edition, 1962.

[5]   Adams, E.P., "Electrical Distributions On Circular Cylinders," *Proc. Am. Phil. Soc. I: Vol. 75*, 1935, *II: Vol. 76*, 1936.

[6]   Sommer, F., "The Calculation of Capacities in Cables With Simple Cross Section," *ENT* (1940), Vol. 17, No. 12, pp. 281-294.

[7]   Knol, K.S., and M.J.O. Strutt, "A Process for the Measurement of Admittances in the Decimeter Region," *Physica IX*, No. 6, June 1942, pp. 577-589.

[8]   Magnus, W., and F. Oberhettinger, "The Calculation of the Characteristic Impedance of a Stripline With Circular or Rectangular Cross Section of the Outer Conductor," *Archiv fur Elektrotechnik*, Vol. 37 (1943), No. 8, pp. 380-390.

[9]   Meinke, and Schlyia, "Deci-File," *Telefunken Report*, Oct.-Nov., 1944.

[10]   Buchholz, H.B., "Calculation of Characteristic Impedance and Damping of High Frequency Transmission Lines from the Field Pattern of the Perfect Conductor," *Archiv fur Elektrotechnik*, I, II, 1948, pp. 79-100, III, 1948, pp. 202-215.

[11]   Oberhettinger, F., and W. Magnus, "Application of Ellip-
        tical Functions in Physics and Engineering," Springer-
        Verlag, 1949.

[12]   Meinke, H.H., "The High Frequency Behavior of the
        Curved Homogeneous Transmission Line," *AEU*, Vol. 5,
        March 1951, No. 3, pp. 106-112.

[13]   Borgnis, F., "The Significance of the Transmission Line
        Equations and of the Characteristic Impedance for Arbi-
        trary Wave Types on Cylindrical Lines," *AEU*, Vol. 5,
        April 1951, No. 4, pp. 181-189.

[14]   Assadourian, F. and E. Rimai "Simplified Theorie of
        Microstrip Transmission Systems," *Proceedings IRE*,
        1952, pp. 1651-1657.

[15]   Morgan, S.P., "Mathematical Theory of Laminated Trans-
        mission Lines," Parts 1, 2, *Bell Syst. Techn.*, J. 31 (1952),
        S. 883-949, 1121-1206.

[16]   Schelkunoff, S.A. and H.T. Friis, *Antennas, Theory and
        Practice*, Wiley, New York, 1952.

[17]   Rösch, H., "The Electrical Field of Conductors Running
        Parallel to Conductive Planes Colliding at a Specific
        Angle," *AEU*, Vol. 8, May 1954, No. 5, pp. 229-237.

[18]   Pease, R.L., and C.R. Mingins, "A Universal Approximate
        Formula for Characteristic Impedance of Strip Trans-
        mission Lines with Rectangular Inner Conductor," *IRE
        Trans. MTT-3*, March 1955, pp. 144-148.

[19]   Oliner, A.A., "Equivalent Circuits For Discontinuities In
        Balanced Strip Transmission Line," *MTT*, 1955, March,
        pp. 134-143.

[20]   Barrett, R.M., "Microwave Printed Circuits — A Historical
        Survey," *MTT-3*, March 1955, pp. 1-9.

[21]   Torgow, E.N., and J.W.E. Griemsmann, "Miniature Strip
        Transmission Line For Microwave Applications," *MTT-3*,
        March 1955, pp. 57-64.

[22]   Zublin, K.E., "Strip Type Components For 2000 Mega-
        cycle Receiver Head — End," *MTT-3*, 1955, pp. 65-74.

[23]   Cohn, S.B., "Problems in Strip Transmission Lines,"
       MTT-3, 1955, pp. 119-126.

[24]   Park, D., "Planar Transmission Lines," *MTT-3*, April 1955,
       pp. 8-12, *MTT-3*, October 1955, pp. 7-10.

[25]   Park, D., "Addendum to Planar Transmission Lines I,"
       MTT, January 1957, pp. 75.

[26]   Hayt, W.H., "Potential Solution of a Homogeneous Strip-
       Line of Finite Width," *MTT-3*, July 1955, pp. 16-18.

[27]   Cohn, S.B., "Shielded Coupled − Strip Transmission
       Line," *MTT-3*, October 1955, pp. 29-38.

[28]   Dahlman, B.A., "A Double − Ground − Plane Strip-Line
       System for Microwaves," *MTT-3*, October 1955, pp. 52-57.

[29]   Bates, R.H.T., "The Characteristic Impedance of the
       Shielded Slab Line," *IRE Trans. MTT-4*, January 1956,
       pp. 28-33.

[30]   Park, D., "Planar Transmission Lines III," *IRE Trans. Vol.
       MTT-4*, April 1956, pp. 130.

[31]   Giger, A.J., "Planar Transmission Lines II," *MTT*, July
       1956, pp. 184.

[32]   Chisholm, R.M., "The Characteristic Impedance of Trough
       and Slab Lines," *MTT*, July 1956, pp. 166-172.

[33]   Butcher, P.N., "The Coupling Impedance of Tape Struc-
       tures," *Proc. IEE*, Vol. 104, March 1957, pp. 177-187.

[34]   Horgan, J.D., "Coupled Strip Transmission Lines with
       Rectangular Inner Conductors," *MTT*, April 1957, pp.
       92-99.

[35]   Packard, K.S., and D. Park, "Planar Transmission Lines,"
       Part III, IV, *MTT*, April 1957, p. 163.

[36]   Packard, K.S., "Optimum Impedance and Dimensions for
       Strip Transmission Line," *MTT*, October 1957, pp. 244-
       247.

[37] Hachemeister, C.A., "The Impedances and Fields of Some
      TEM-Mode Transmission Lines," *Research Report R-623-
      57, PIB-551* for Air Force Cambridge Research Center,
      April 16, 1958, Polytechnic Institute of Brooklyn, Micro-
      wave Research Institute, Electrophysics Group.

[38] Smolarska, J., "Characteristic Impedances of the Slotted
      Coaxial Line," *MTT*, April 1958, pp. 161-166.

[39] Oehrl, W., G. Seeger, and H.G. Stäblein, "General Con-
      siderations of the Theory of Multiple Lines," *AEU*, Vol.
      12, June 1958, No. 6, pp. 245-250.

[40] De Buda, R.G., "A Method of Calculating the Character-
      istic Impedance of a Strip Transmission Line to a Given
      Degree of Accuracy," *MTT*, October 1958, pp. 440-446.

[41] Harvey, A.F., "Parallel-Plate Transmission Systems for
      Microwave Frequencies," *Proc. IEE*, Vol. 106, March 1959,
      pp. 129-140.

[42] Lauterjung, K., "Study of Symmetrical High Frequency
      Transmission Lines," *AEU*, Vol. 14, January 1960, No. 1,
      pp. 26-36.

[43] Kogo, Hiroshi, "A Study of Multielement Transmission
      Lines," *MTT*, March 1960, pp. 136-142.

[44] Owyang, G.W., "Complementarity in the Study of Trans-
      mission Lines," *MTT*, March 1960, pp. 172-181.

[45] Cohn, S.B., "A Reappraisal of Strip Transmission Line,"
      Microwave J., Vol. 3, pp. 17-27, March 1960 (Litera-
      turübersicht).

[46] Getsinger, W.J., "Analysis of Certain Transmission − Line
      Networks in the Time Domain," *MTT*, May 1960, pp.
      301-309.

[47] Altschuler, H.M., and A.A. Oliner, "Discontinuities in the
      Center Conductor of Symmetric Strip Transmission Line,"
      MTT, May 1960, pp. 328-339.

[48] Tsung − Shan Chen, "Determination of the Capacitance,
      Inductance, and Characteristic Impedance of Rectangular
      Lines," *MTT-8*, September 1960, pp. 510-519.

[49]   Cohn, S.B., "Thickness Corrections for Capacitive Ob-
       stacles and Strip Conductors," *MTT*, November 1960,
       pp. 638-644.

[50]   Cohn, S.B., "Characteristic Impedances of Broadside —
       Coupled Strip Transmission Lines," *MTT*, November 1960,
       pp. 633-637.

[51]   Geschwinde, H., and W. Frank, *Striplines*, Winter, Verlag,
       Fussen, 1960.

[52]   Garver, R.V., "Z_0 of Rectangular Coax," *MTT*, May 1961,
       pp. 262-263.

[53]   Getsinger, W.J., "A Coupled Strip — Line Configuration
       Using Printed — Circuit Construction that Allows Very
       Close Coupling," *MTT*, November 1961, pp. 535-544.

[54]   Buschbeck, "The Characteristic Impedance of Cylindrical
       Conductors Opposed to Various Systems of Shielding
       Walls," *Telefunken-Zeitung*, 1961, No. 131, pp. 69-76.

[55]   Getsinger, W.J., "Coupled Rectangular Bars Between
       Parallel Plates," *MTT*, January 1962, pp. 65-72.

[56]   Gaal, E., "Current Distribution on a Stripline," *Acta
       Technica*, 1962, pp. 387-397.

[57]   McQuillan, J.D.R., "The Design Problems of a Megabit
       Storage Matrix for Use in a High-Speed Computer," *IRE
       Transact. E.C.*, June 1962, pp. 390-404.

[58]   Kessler, A., A. Vlcek, and O. Zinke, "Methods for the
       Determination of Capacities With Special Consideration of
       the Partial Surface Method," *AEU*, Vol. 16, August 1962,
       No. 8, pp. 365-380.

[59]   Buschbeck, "The Characteristic Impedance of an Eccentric
       Cylindrical Conductor Between Parallel Planes, or a Cylin-
       drical Conductor in a Central Plane of a Rectangular Box,"
       Telefunken-Zeitung, 1962, No. 138, pp. 361-364.

[60]   Nicolai, K., "Tables of the Complete Elliptical Integrals of
       the First Type," *Telefunken*, 1962.

[61] Nicolai, K., "Tables of the Characteristic Impedance of the High-Q Triplate," *Telefunken*, 1962.

[62] Piefke, G., "Damping and Distortion of a Pulse on a Stripline with Magnetic Interlayer," *AEU*, Vol. 17, April 1963, No. 4, pp. 153-162.

[63] Jutzi, W., "The Effect of Eddy Currents on the Specific Inductance and Resistance of Unsymmetrical Parallel Striplines," *AEU*, Vol. 17 (1963), September, No. 9, pp. 420-428.

[64] Seshagiri, N., "A Nonuniform Coaxial Line with an Isoperimetric Sheath Deformation," *MTT*, November 1963, pp. 478-486.

[65] Seshagiri, N., "A Uniform Coaxial Line with an Elliptic − Circular Cross Section," *MTT*, November 1963, pp. 549-551.

[66] Green, H.E., "The Characteristic Impedance of Square Coaxial Line," *MTT*, November 1963, pp. 554-555.

[67] Rauskolb, R.F., and G.F. Landvogt, "The Most Favorable Cross Section Dimensions of Double Transmission Lines and Symmetrical Wide Strip Cables with Respect to Capacity, Electric Strength, and Damping," *AEU*, Vol. 18, January 1964, No. 1, pp. 67-76.

[68] Wheeler, H.A., "Transmission − Line Properties of Parallel Wide Strips by a Conformal − Mapping Approximation," *MTT*, May 1964, pp. 280-289.

[69] Conning, S.W., "The Characteristic Impedance of Square Coaxial Line," *MTT*, July 1964, p. 468.

[70] Cristal, E.G., "Coupled Circular Cylindrical Rods Between Parallel Ground Planes," *MTT*, July 1964, pp. 428-439.

[71] Cruzan, O.R., and R.V. Garver, "Characteristic Impedance of Rectangular Coaxial Transmission Lines," *MTT*, September 1964, pp. 488-495.

[72] Matthaei, G.L., L. Young, and E.M.T. Jones, *Microwave Filters, Impedance-Matching Networks, And Coupling Structures*, McGraw-Hill Book Company, New York, 1964.

[73]   Nicolai, K., "Tables of the Characteristic Impedances of a Triplate with Rectangular Internal Conductor," *Telefunken*, 1964.

[74]   Duncan, J.W., "Characteristic Impedances of Multiconductor Strip Transmission Lines," *MTT-13*, 1965, January, pp. 107-118.

[75]   Green, H.E., "The Characteristic Impedance and Velocity Ratio of Dielectric − Supported Strip Line," *MTT-13*, January 1965, pp. 141.

[76]   Correction on: Cristal, *MTT-13*, July 1964, pp. 428-439, "Coupled Circular Cylindrical Rods Between Parallel Ground Planes," *MTT-13*, January 1965, pp. 141.

[77]   Vadopalas, P., and E.G. Cristal, "Coupled Rods Between Ground Planes," *MTT-13*, March 1965, pp. 254-255.

[78]   Wheeler, H.A., "Transmission − Line Properties of Parallel Strips Separated by a Dielectric Sheet," *MTT-13*, March 1965, pp. 172-185.

[79]   Guckel, H., "Characteristic Impedances of Generalized Rectangular Transmission Lines," *MTT-13*, May 1965, pp. 270-274.

[80]   Jutzi, W., "The Magnetic Pulse Field of an Unsymmetrical Parallel Stripline," *AEU*, No. 19 (1965), pp. 119-125.

[81]   Burton, R.W., and R.W.P. King, "An Experimental Investigation of a Two-Slot Transmission Line on Nonplanar Surfaces," *MTT-13*, May 1965, pp. 303-306.

[82]   Cruz, J.E., and R.L. Brooke, "A Variable Characteristic Impedance Coaxial Line," *MTT-13*, July 1965, pp. 477-478.

[83]   Green, H.E., "The Numerical Solution of Some Important Transmission − Line Problems," *MTT-13*, September 1965, pp. 676-692.

[84]   Hyltin, T.M., "Microstrip Transmission on Semiconductor Dielectrics," *MTT-13*, November 1965, pp. 777-781.

[85]   Schneider, M.V., "Computation of Impedance and Attenu-
        ation of TEM-Lines by Finite Difference Methods,"
        MTT-13, November 1965, pp. 793-800.

[86]   Feller, A., H.R. Kaupp, and J.J. Digiacomo, "Crosstalk and
        Reflections in High-Speed Digital Systems," *Afips Conf.
        Proc.*, Vol. 27, Part I, 1965, Fall Joint Computer Confer-
        ence, pp. 511-525.

[87]   Baier, W., "Waves in Hollow Conductors of Very General
        Cross Sectional Shape," Thesis, TH Munchen, 1965.

[88]   Kurz, G., "Triplate Line, Part I: Theoretical Considera-
        tions," *Telefunken*, 1965.

[89]   Shelton, J.P., Jr., "Impedances of Offset Parallel-Coupled
        Strip Transmission Lines," *MTT-14*, January 1966, pp.
        7-15. Correction in: *MTT-14*, May 1966, p. 249.

[90]   Ravi, C.G., and G.G. Koerber, "Effects of a Keeper on
        Thin Film Magnetic Bits," *IBM Journal of Research and
        Development*, Vol. 10, No. 2, March 1966, pp. 130-134.

[91]   Palocz, J., "The Integral Equation Approach to Currents
        and Fields in Plane Parallel Transmission Lines," *Journal
        of Mathematics and Mechanics*, Vol. 15, No. 4, 1966, pp.
        541-559.

[92]   Caulton, M., J.J. Hughes, and H. Sobol, "Measurements on
        the Properties of Microstrip Transmission Lines for Micro-
        wave Integrated Circuits," *RCA Review*, Vol. 27, 1966,
        September, pp. 377-391.

[93]   Yamamoto, S., T. Azakami, and K. Itakura, "Coupled Strip
        Transmission Line With Three Center Conductors,"
        MTT-14, October 1966, pp. 446-461.

[94]   Carson, C.T., and G.K. Cambrell, "Upper and Lower
        Bounds on the Characteristic Impedance of TEM Mode
        Transmission Lines," *MTT-14*, October 1966, pp. 497-498.

[95]   Cremosnik, G., "Contributions to the Solution of the
        Static Fields of Line Charges Between Two Wedge-Shaped
        Planes," *AEU*, Vol. 20, October 1966, No. 10, pp. 579-
        583.

[96] Pfügel, D., "The Partial Surface Method for Determination
      of the Capacity of Arbitrary Conductors," *Zeitschrift fur
      angewandete Physik 23*, Vol., No. 2, 1967, pp. 89-94.

[97] Macario, R.C.V., "Measurement and Calculation for
      Magnetically Coated Strip Transmission Lines," *IEEE
      Transactions on Magnetics*, Vol. MAG-3, No. 1, March
      1967, pp. 6-9.

[98] Bräckelmann, W., D. Landmann, and W. Schlosser, "The
      Limiting Frequencies of Higher Standing Waves in Strip-
      lines," *AEU 21*, 1967, No. 3, pp. 112-120.

[99] Kaupp, H.R., "Characteristics of Microstrip Transmission
      Lines," *IEEE Transactions*, E.C. Vol. EC-16, No. 2, April
      1967, pp. 185-193.

[100] Jutzi, W., "The Magnetic Field of an Unsymmetrical
      Parallel Stripline with Magnetic Ground," *AEU 21* (1967),
      No. 4, pp. 190-196.

[101] Swift, J., "Strip Transmission Lines," *Electronic Engi-
      neering*, August 1967, pp. 490-494.

[102] Guckel, H., P.A. Brennan, and J. Palocz, "A Parallel-plate
      Waveguide Approach to Microminiaturized, Planar Trans-
      mission Lines for Integrated Circuits," *IEEE Transaction
      MTT*, Vol. MTT-15, No. 8, August 1967, pp. 468-476.

[103] Mohr, R.J., "Coupling Between Open and Shielded Wire
      Lines Over a Ground Plane," *IEEE Transactions on Electro-
      magnetic Compatibility*, Vol. EMC-9, No. 2, September
      1967, pp. 34-45.

[104] Schwarzmann, A., "Microstrip Plus Equations Adds up to
      Fast Designs," *Electronics*, October 2, 1967, pp. 109-112.

[105] Hilberg, W., "The Possibility of Replacing Certain Charac-
      teristic Impedance Formulas Which Contain Elliptical In-
      tegrals by Approximation Formulas of Arbitrarily High
      Selectable Accuracy," *AEU 21* (1967), No. 11, pp. 603-
      616.

[106] Berle, F.J., "Study of the Partial Surface Method in the Calculation of the Capacity of Slotted Coaxial Transmission Lines," *Frequenz, Band 21*, November 1967, No. 10, pp. 333-343.

[107] Kurz, G., "Triplate Line, Part II: Line Elements," *AEG-Telefunken*, 1967.

[108] Berle, F.J., "Application of the Partial Surface Method in the Calculation of the Partial Capacities of a Slotted Coaxial Line, with Consideration of Wall Thicknesses Above an Ideal Conducting Plane," *AEU 21*, 1967, No. 12, pp. 674-676.

[109] Bräckelmann, W., "Wave Types in the Stripline With Rectangular Shield," *AEU 21*, 1967, No. 12, pp. 641-648.

[110] Hilberg, W., "The Decoupling of Lines Consisting of Two Conductors," *AEU 22* (1968), No. 1, 39-45.

[111] Ciganek, L., "Complete Solution of Some Important Cases of Two Dimensional Magnetic Fields with Rectangular Boundaries," *Acta Technica CSAV*, 1968, No. 1, pp. 68-99.

[112] Hilberg, W., "A Simple and Good Approximation for the Characteristic Impedance of Parallel Striplines," *AEU 22* (1968), No. 3, 122-126.

[113] Baier, W., "Wave Types in Lines Consisting of Conductors with Rectangular Cross Section," *AEU* (1968), No. 4, pp. 179-185.

[114] Yamashita, E., and R. Mittra, "Variational Method for the Analysis of Microstrip Lines," *IEEE MTT*, Vol. 16, No. 4, April 1968, S. 251-256.

[115] Jlenburg, W.R., and R. Pregla, "The Limiting Frequencies of Higher Wave Forms in A System with Several Striplines," *AEU* (1968), No. 5, pp. 230-238.

[116] Krage, M.K., and G.J. Haddad, "The Characteristic Impedance and Coupling Coefficient of Coupled Rectangular Strips in a Waveguide," *IEEE MTT*, Vol. 16, No. 5, May 1968, S. 302-307.

[117] Cristal, E.G., "Data for Partially Decoupled Round Rods Between Parallel Ground Planes," *IEEE MTT*, Vol. 16, No. 5, May 1968, 311-314.

[118] Pucel, R.A., D.J. Masse, and C.P. Hartwig, "Losses in Microstrip," *IEEE Transactions MTT*, Vol. MTT-16, No. 6, June 1968, pp. 342-350.

[119] Bräckelmann, W., "Capacities and Inductances of Coupled Striplines," *AEU*, Vol. 22, July 1968, No. 7, pp. 313-321.

[120] Clemm, H.L., "Specific Capacity and Characteristic Impedance of Unsymmetrical Striplines," *Frequenz 22* (1968), 7, pp. 196-201.

[121] Stinehelfer, H.E., "An Accurate Calculation of Uniform Microstrip Transmission Lines," *IEEE Transactions MTT*, Vol. MTT-16, No. 7, July 1968, pp. 439-444.

[122] Pregla, R., and W. Schlosser, "Waveguide Modes in Dielectrically Supported Strip Lines," *AEU* (1968), H. 8, pp. 379-386.

[123] Yamashita, E., "Variational Method for the Analysis of Microstrip-like Transmission Lines," *IEEE MTT*, Vol. 16, No. 8, August 1968, pp. 529-535.

[124] Whiting, K.B., "A Treatment for Boundary Singularities in Finite Difference Solutions of Laplace's Equation," *IEEE MTT*, Vol. 16, No. 10, October 1968, pp. 889-891.

[125] Kammler, D.W., "Calculation of Characteristic Admittances and Coupling Coefficients for Strip Transmission Lines," *IEEE, MTT*, Vol. 16, No. 11, November 1968, pp. 925-937.

[126] Sigg, H., and M.J.O. Strutt, "The Calculation of the High Frequency Parameters of Coupled Parallel Transmission Lines with the Use of an Analogy Network," *AEU* (1968), No. 11, pp. 543-552.

[127] Beaubien, M.J., and A. Wexler, "An Accurate Finite-difference Method for Higher Order Waveguide Modes," *IEEE, MTT*, Vol. 16, No. 12, December 1968, pp. 1007-1017.

[128] Bryant, Th. G., and J.A. Weiss, "Parameters of Microstrip Transmission Lines and of Coupled Pairs of Microstrip Lines," *IEEE, MTT*, Vol. 16, No. 12, December 1968, pp. 1021-1027.

[129] Pucel, R.A., Correction to "Losses in Microstrip," *IEEE, MTT*, Vol. 16, No. 12, December 1968, pp. 1064.

[130] Yohan, Cho, and John Connolly, "System Packaging for Sub-nanosecond Micrologic Circuits," *Mikroelektronik München*, November 1968.

[131] Sinnott, D.H., "The Use of Interpolation in Improving Finite Difference Solutions of TEM Mode Structures," *IEEE MTT*, Vol. 17, No. 1, January 1969, pp. 20-28.

[132] Schwarzer, H., "The Magnetic Field Strengths in Symmetrical Striplines for Magnetic Wire Storage," *AEU 23*, 1969, No. 2, pp. 87-93.

[133] Campbell, J.J., and W.R. Jones, "Impedance Characteristics of a Class of Multiconductor Transmission Lines," *MTT*, Vol. 17, No. 2, February 1969, pp. 101-107.

[134] Hilberg, W., "From Approximations to Exact Relations for Characteristic Impedances," *IEEE Transactions MTT*, Vol. 17, No. 5, May 1969, pp. 259-265.

[135] Kamman, K., S. Stief, and R. Wohlleben, "Characteristic Impedance of Complicated Cross Section Forms by the Method of Approximative Conformal Mapping of Doubly Correlated Regions," *AEU* (1969), No. 5, pp. 221-228.

[136] Baier, W., "Calculation of Limiting Frequencies of Inhomogeneously Filled Hollow Rectangular Conductors and the Field Conditions Appearing at these Frequencies," *AEU* (1969), No. 5, pp. 237-241.

[137] Clemm, H.L., "Calculation of Capacity and Characteristic Impedance of the Stripline on a Dielectric Carrier (Microstrip) with the Use of the Partial Surface Method," *Frequenz 23* (1969), 5, pp. 143-151.

[138] Mahr, H., "A Contribution to the Theory of the Coaxial Line Excited in the Ground Wave Method," *Der Fernmelde-Ingenieur*, 23, Jahrg, 1969, No. 5, No. 6, No. 7.

[139] Zysman, G.I. and D. Varon, "Wave Propagation in Microstrip Transmission Lines," *Int. Microwave Symp.*, Dallas, May 5-7, 1969.

[140] Hilberg, W., "Elementary Treatment of the Overcoupling of Pulses and Sine Waves Between Parallel Lines," *NTZ*, 22, January, June, 1969, No. 6, pp. 368-373.

[141] Hilberg, W., "Possibilities and Limitations of an Elementary Theory of Overcoupling of Pulses and Sine Waves Between Parallel Transmission Lines," *NTZ-Report*, No. 4, VDE-Verlag Berlin, 1969.

[142] Schneider, M.V., "Microstrip Lines for Microwave Integrated Circuits," *Bell System Technical Journal*, May-June 1969, pp. 1421-1444.

[143] Gupta, R.R., "Accurate Impedance Determination of Coupled TEM Conductors," *MTT*, Vol. 17, No. 8, August 1969, pp. 479-489.

[144] Yamashita, E., and K. Atsuki, "Design of Transmission – Line Dimensions for a Given Characteristic Impedance," *MTT*, Vol. 17, No. 8, August 1969, pp. 638-639.

[145] Sinnott, D.H., "Calculation of TEM Transmission-Line Parameters by Finite-Difference Computation of Electric Flux," *IEEE MTT*, Vol. 17, No. 8, August 1969, pp. 634-637.

[146] Sinnott, D.H., G.K. Cambrell, C.T. Carson, and H.E. Green, "The Finite Difference Solution of Microwave Circuit Problems," *IEEE MTT*, Vol. 17, No. 8, August 1969, pp. 464-478.

[147] Chestnut, P.C., "On Determining the Capacitances of Shielded Multiconductor Transmission Lines," *MTT*, Vol. 17, No. 10, October 1969, pp. 753-759.

[148] Zysman, G.J., "Coupled Transmission Line Networks in an Inhomogeneous Dielectric Medium," *IEEE MTT*, Vol. 17, No. 10, October 1969, pp. 753-759.

[149] Kraus, A., and H.J. Renger, "The Calculation, Measurement, and Representation of Families of Curves of the Intrinsic Capacities of Multiple Transmission Lines," *AEU* (1969), No. 10, pp. 489-501.

[150] Wohak, K., "Approximate Calculation of the Inductance of Thin Layered Paths," *Frequenz 23* (1969), 12, pp. 359-364.

[151] Grivet, P., *The Physics of Transmission Lines at High and Very High Frequencies*, Academic Press, London 1970.

[152] Weeks, W.T., "Calculation of Coefficients of Capacitance of Multiconductor Transmission Lines in the Presence of a Dielectric Interface," *IEEE MTT*, Vol. 18, No. 1, January 1970, pp. 35-43.

[153] Judd, S.V., I. Whiteley, R.J. Clowes, and D.C. Rickard, "An Analytical Method for Calculating Microstrip Transmission Line Parameters," *IEEE MTT*, Vol. 18, No. 2, February 1970, pp. 78-87.

[154] Mittra, R., and T. Jtoh, "Carge and Potential Distribution in Shielded Striplines," *IEEE MTT*, Vol. 18, No. 3, March 1970, pp. 149-156.

[155] Sato, R., and E.G. Christal, "Simplified Analysis of Coupled Transmission-line Networks," *IEEE MTT*, Vol. 18, No. 3, March 1970, pp. 122-131.

[156] Gish, D.L., and O. Graham, "Characteristic Impedance and Phase Velocity of a Dielectric-Supported Air Strip Transmission Line with Side Walls," *IEEE MTT*, Vol. 18, No. 3, March 1970, pp. 131-148.

[157] Krage, M.K., and G.J. Haddad, "Characteristics of Coupled Microstrip Transmission Lines, I: Coupled-mode Formulation of Inhomogenious Lines, II: Evaluation of Coupled-line Parameters," *IEEE MTT*, Vol. 18, No. 4, April 1970, pp. 217-228.

[158] Hilberg, W., "Approximations for the Elliptical Integral Function K/K' and Recursions for the Optional Improvement of Their Accuracy, Particularly for the Calculation

of the Characteristic Impedance," *AFE*, 53, Vol., 1970, No. 5, pp. 290-298.

[159] Yamashita, E., "Strip Line with Rectangular Outer Conductor and Three Dielectric Layers," *IEEE MTT*, Vol. 18, No. 5, May 1970, pp. 238-244.

[160] Wen, Ch. P., "Coplanar-Waveguide Directional Couplers," *IEEE Transactions MTT*, Vol. 18, No. 6, June 1970, pp. 318-322.

[161] Burmester, A., "The Calculation of Capacities in Cables with Simple Cross Section with Consideration of Non-Homogeneous Insulation," *AEU* (1970), No. 9, pp. 395-400.

[162] Hilberg, W., "Stringent Calculation of the Characteristic Impedance of Parallel Striplines and a Comparison with Approximations," *AFE*, 54 (1971), pp. 200-205.

[163] Verma, S.M., and P.E. Gamble, "Characteristic Impedance of Flat Unshielded Multi-Conductor Cables," *Proc. 19th Int. Wire and Cable Symposium*, Atlantic City, NJ, 1-3 December 1970.

[164] Bossi, D.F., "Digital Speeds Versus the Flat Cable Transmission Line," *Proc. 19th Int. Wire and Cable Symposium*, Atlantic City, NJ, 1-3 December 1970.

www.ingramcontent.com/pod-product-compliance
Lightning Source LLC
Chambersburg PA
CBHW021431180326
41458CB00001B/220